D0844217

The Energy Crisis

The Energy Crisis

Lawrence Rocks
and
Richard P. Runyon

CROWN PUBLISHERS, INC., NEW YORK

Library of Congress Catalog Card Number: 72-89410
ISBN: 0-517-501643
ISBN: 0-517-501651

Printed in the United States of America
Published simultaneously in Canada by General Publishing Company Limited

To
Marlene and Lois

Acknowledgments

In writing this book, we have relied heavily upon the pioneering efforts of many men of science and industry who have sighted the impending energy crisis and have contributed much of the basic data with which we have worked. Their names and their contributions are cited frequently throughout this book: M. King Hubbert, world-famed geologist; Glenn T. Seaborg, former chairman of the Atomic Energy Commission; William R. Corliss, science writer; Ralph Lapp, scientist-writer; Richard A. Rice, Professor of Transportation at Carnegie-Mellon University; and Lawrence M. Lidsky, Professor of Nuclear Engineering at MIT. Although we have drawn from these and other sources, we take full responsibility for the interpretations we have made.

We should also like to acknowledge the dedicated work of Dorothy Loose, Rosemary Kopczynski, and Mabel Perham who cheerfully typed the various drafts of the manuscript. Willard Warwick prepared all the graphic illustrations appearing in the book.

A special word of appreciation is due Paul Nadan, our editor, for his overall guidance in the preparation of this book.

Finally, we wish to acknowedge the Research Center of C. W. Post College for providing a grant to assist in the preparation of the manuscript.

Contents

Authors' Preface

The Bible records that God, the Creator, fashioned man in his own image. On such a biblical basis man could be seen as summoned to a special mission, that of a creative career on earth. However, even putting aside the biblical interpretation, one can hardly deny man's enormous creative talents as revealed by his works.

Down through the millennia, the painfully accumulated knowledge of mankind grew, slowly at first, then more rapidly – steadily empowering man to vie with his environment, even alter it, until with ever-increasing speed, he climbed from the physical exertions in prehistory to the lofty and powerful industrial stance of our time. It has been a dramatic ascent, accelerated in recent years to the heights of mass production and mass communication. In the struggle, muscle gave way to simple tools; these in turn were superseded by machines. Each progression was a consequence of man's having learned to exploit or harness yet another natural resource or phenomenon. By the gift of creativity, we have been carried to a new environment – an industrial environment of our own making.

But our ascent has been marred. Our progress has brought various threats to the verge of climaxing: many key materials and energy resources are near exhaustion. In fact, energy resources are so limited that not only can't the present living standard of the United States be available to all mankind, it is not likely to continue in this country beyond the present decade. But population keeps growing and living-standard desires keep rising. Around the world, people dream of participating in the latest scientific discoveries through a rising living standard – all the while oblivious to the true magnitude of the energy requirements necessary to utilize such new knowledge.

The thesis of this book is that the most profound issue we face today is an impending power shortage. Most other environmental problems are theoretically solvable: pollution can be largely abated by the same science that created industry, and substantial percentages of materials can be recycled, and thereby held in perpetuity. But energy cannot be recycled! In fact, our energy resources are being rapidly depleted. In the ensuing decade the deepening stresses of a gathering power shortage will clearly reveal the great role that power plays in our lives.

Our energy deployment capabilities and their consequences will supersede all other environmental, economic, and political issues before this decade has passed. The role of power in human affairs will become the greatest influence in determing living-standard horizons.

Power is such a pervasive element in our everyday living that we tend to take it for granted. It is modern man's slave. If we continue to treat it with the neglect and disdain usually accorded to slaves, it will die. With the death of our mute and uncomplaining servant, we shall also die. Reflect upon the role of power in your life.

Your body runs on power. This power is derived from the food you take in daily. Withhold food from the body and the body ceases to function.

Your household runs on power. From the clock radio that summons you from sleep to the electric blanket that warms you throughout the cold winter's night, your household consumes power in much the same way our bodies consume food to maintain life. How do you heat your home in the winter? Coal? Oil? Natural gas? Electricity? How do you cool it in the summer? Air conditioning? Electric fans? How do you keep perishable food safe for consumption? How do you see at night? How do you mix cake batter, cook a meal, wash and dry your clothing? Withdraw power from your household and it ceases to function.

Probably you have experienced a temporary loss of power from, for example, the various blackouts in the Eastern United States, power lines knocked down during a storm, and so on. You have found your routines completely interrupted. You have have even found these episodic losses of power somewhat pleasurable since they have momentarily transported you from the tedium of daily routines. The pleasure is short lived as the worries begin to accumulate; the hazards of travel and how to get home, food spoilage, the loss of work, and so on. What if the power loss were to become protracted and chronic?

Your industrial life-support systems run on power. Why are the farms of the nation vastly more productive than they were fifty years ago? Power. How are the many necessities of life manufactured? Power. The amenities? Power. How do you and millions of Americans get to and return from work daily? How do the many products of industry get to your home? Withdraw the power from agriculture and industry and our industrial life-support systems cease to function.

Clearly, the power issue, the impending decline in power availability, will affect all of us severely and immediately. We are facing constraints that could drastically lower our living standards, and bring us to the sad condition of holding ration cards for gasoline. We face voltage reductions in summer, with restricted use of air conditioners, and gas shortages in winter, with less heating on cold days. And that is just a start.

America faces an immediate, severe power shortage during the 1970s and 1980s. We face this disaster without any coherent national design for meaningful power procurement.

An energy gap is opening up in America sufficiently wide to cause our total industrial collapse in one decade. It is doubtful that the gap can be closed, for we have let pass our most timely chances for beginning the necessary power supply changes that are needed.

During the next two decades, severe oil and gas shortages are inevitable! We shall be powerless to infuse energy sources on a sufficiently massive scale to meet the demands of our industrial life-support systems. This conclusion holds true for synthetic fuels, that is, oil and gas synthesized from coal, as well as for atomic energy utilizing uranium.

Our task must be to raise the energy issue from relative obscurity to the dominant issue of our times. We must then conceive a meaningful energy program as quickly as possible. Our energy posture is precarious. In addition to our severe short supply of gas and oil, in two or three decades there will be no more repositories of these fuels; we cannot produce "synthetic fuels" from coal at a fast enough rate to maintain America's self-sufficiency in energy; atomic energy (uranium) is about five decades from supplying our electrical needs; we have not yet committed the nation to a massive research effort to develop fusion power; and we have not resolved the satellite needs of an electrical age. Power expressed as electricity is man's only energy mode after the expiration of the fossil fuels. We are unprepared for an electrical age.

We are running out of power and have no plans that can close the energy gap in time to avert a shutdown of American industry, and with that, a collapse of our way of life and loss of our national independence.

Long before we reach our last drop of oil or our last pound of coal, we need alternative power systems in operation – alternative power systems that have been conceived, tested, perfected, and deployed. Development of such systems requires many years and large infusions of money.

If a power shortage is real, why isn't there more publicity? Power has not been an easily understood subject. It has always demanded the use of numbers for real understanding, very much like the population explosion. To appraise the power situation accurately, one must be comfortable with the large numbers that describe energy banks and power requirements, and future power options.

Also there are the time scales. One must think in decades and centuries, obviously a major portion of and exceeding one's lifetime. Such chronological horizons are difficult to grasp; they are easily misapprehended from a motivation to avoid thinking about the unacceptable, overbearing problems of today.

And the time dimensions are so great as to evoke only an emotional, subjective, limited response. How is any citizen to react, challenged and buffeted from a hundred directions every minute of his waking life in our fast-paced society, to the announcement that we will see a water shortage crisis in the year 2000? "We've had them before," will be the response. A copper shortage in 1990 and curtailment of all electricity? "We'll think of something by then," will be the response.

And so, we may have the knowledge of an impending power default, *but not the sense of urgency to avert it.*

Intelligent life has two power alternatives for the very distant future: (1) to develop a long-lived source of power or (2) to live in balance with the finite power sources. Since the latter can be accomplished only by a drastic reduction in our population, living standard, or both, there is really only one alternative – fusion power. It represents the best prospect for a long-term, accessible, abundant, and inexpensive energy source. If we hurry now, we may possibly lessen the onrushing power default and ease our entrance into the age of fusion power. We may yet tap the enormous potential that resides in the atom. To date, we have only glimpsed its possibilities in the awesome energy released by the detonation of the hydrogen bomb. If we do not have a meaningful power policy by 1976, then certainly we shall have passed the point of no return. We shall have a great deal to say about fusion power in later chapters.

Estimates of energy resources throughout this book are from the most reliable sources available, many from the National Academy of Sciences. Nevertheless, estimating a resource is an extremely complex affair – as much art as science. Various scientists, employing different methods of estimation, do not necessarily arrive at the same values. Moreover, definitions vary concerning the amounts that can be economically extracted. Thus, we find one source estimating the total world reserves of copper at 308 million tons whereas the National Academy of Sciences estimates the reserves at about 200 million tons.[1] In truth, much of this copper may never be recovered, either because it is energetically too expensive or financially too expensive to extract. Nevertheless, for wide differences in estimates, the figure used will be a range of time during which the resource may be expected to be exhausted.

Similarly, there is not always agreement concerning the rates of utilization of a resource. At times, then, various "if" statements will be used. That is, "if" a resource is in a given supply and "if" its rate of utilization continues at a certain level, we may expect it to be exhausted by such and such a date.

Comprehension of the "rate" problem is fundamental to understanding the message of this book. Why is "rate" so pivotal? Stated simply, as we exhaust the fossil fuels in their more accessible locations we must begin looking for and developing these resources in more remote and inaccessible locations (e.g., Alaska, the Arctic region, etc.). It is a simple and indisputable fact that it takes more time, effort, and finances to extract and transport fossil fuels from these remote locations. Or, as we shall see, it also takes much time, effort, and money to convert a more accessible resource, such as coal, into more readily usable forms, such as gas and oil.

Before starting our story, several assumptions must be made clear.

(1) We do not believe there is *any* power-resources strategy that will permit *all* of mankind to enjoy the energy consumption rates that presently prevail in America. To attempt to do so would be suicidal. All energy sources at the delivery end are polluting in some way – air, radioactive, or thermal.[2] The earth is incapable of sustaining the enormous increases that would be necessary to bring the rest of humanity up to our standards of power consumption. In addition, as we shall see, shortages of key resources necessary to deliver power will sharply limit our capacity to expand energy consumption rates.[3]

(2) Restrictions upon future rates of energy consumption are inescapable. There will be many soul-searching, agonizing days ahead as we seek to accomplish these objectives. More than ever before we shall require cool heads as we contemplate the various alternatives: more efficient employment of energy-exercising

devices, voluntary and involuntary controls on the use of these devices, and a gradual and planned reduction in the earth's population, to name a few. Our book is part optimistic, part pessimistic. Our optimism stems from the realization that options are available to us. We would be less than honest, however, if we did not admit to a pervasive sense of pessimism. The present world political and social climate, with man pitted against man and nation against nation, appears to preclude the massive cooperative worldwide effort necessary to keep the lights turned on.

(3) Many of our institutions cannot survive the several power-base transformations facing us during the next five decades. Solutions of political, social, and economic problems, if they are forthcoming, will require a pooling of talents from all walks of life: physical scientists, behavioral and social scientists, economists, political scientists, and the oft-ignored talent that resides in the humanities and the arts.

We shall argue for the pursuit of a long, flexible policy for national and international survival: we must win a larger energy base during the next several decades in order to buy the time and opportunity to develop an atomic-powered industrial environment and the worldwide acceptance of a plan to ultimately lessen world population and resource consumption rates.

Let us develop the basis for recommending this sequence of phases of opposite energy goals.

NOTES

1. Donnella H. Meadows, Dennis L. Meadows, Jorgen Randers, and William W. Behrens III, *The Limits to Growth* (New York: Universe Books, a Potomac Associates, 1972).
2. Ibid., ch. 3.
3. Ibid.

The Accelerating Power Crisis

AMERICA BLACKED OUT

What would the United States be like if we suffered a massive power default, with no viable alternative energy sources? Obviously, we cannot know. But we can make some informed guesses.

First, let us assume that the default will not be abrupt. It will be slow at first, gathering momentum as oil becomes increasingly inaccessible and we expend greater amounts of energy to obtain, refine, and transport it.

At the beginning, the cost of gasoline and oil products will increase. This increase will represent a minor inconvenience to many and a financial hardship for some. Also, during periods of heavy electrical use, power companies will have to reduce voltage to home and industrial users. Annoying and seemingly trivial difficulties will be encountered in the use of home appliances. Some will not function properly under reduced voltage. Others will have their life expectancies shortened because a loss of power will not permit them to operate efficiently.

As the shortage grows more severe, there will be restrictions placed upon the use of luxuries. The use of air conditioners will be prohibited or severely limited. Oil for heat in the home will be cut back.

As the oil shortage deepens and begins to approach crisis levels, concern for loss of luxuries or creature comforts will give way to an enormous preoccupation with obtaining life's essentials. Rationing of gasoline and fuel oil will be mandated. People will be encouraged or required to use mass transportation to get to and from work. They will be exhorted or required to form car pools in locations where mass transportation is not available. Pleasure driving will exist only as fond memories.

Lowered use of the automobile will necessitate drastic cutbacks in automobile manufacturing. Massive layoffs will occur in the auto industry. Other industries, dependent wholly or in part upon the automobile industry, will suffer more extensive layoffs. The steel industry will be incapacitated. A domino effect will ensue throughout the economy.

The shortage of oil will continue to worsen. Congress will enact legislation restricting the use of gasoline to essential uses, priorities will be established, with abuses subject to severe penalties. Food production, as a first priority, will be initially unaffected. However, to get the food from the farm to the market and then to the home requires power. Our ground transportation system, completely dependent upon fossil fuels, will develop, at first, localized and sporadic shortages. The shortages of transportation fuel will grow more general and chronic. Difficulties in transporting necessities to and from the farm will begin to affect farm production. At this point, we will come to comprehend fully the extent to which our incredible agricultural productivity has depended upon this gift from nature – our "oil subsidy."

As our physical systems that produce goods and services become paralyzed we will be forced to take steps to sustain our life-support systems. Decisions will be required that will shortcut our economic and social freedom. Survival will supersede due legal process. Energy rationing, with all its authoritarian consequences, will take precedence over the prerogatives of our once free economic institutions. Our entire network of interrelated institutions will feel the shock waves. Forgotten will be our lofty ideas – our concepts of justice and democracy; distinctions between capitalism and Communism will appear to be hair-splitting academic exercises. We will be constantly and overwhelmingly preoccupied with survival. In the face of the controls and regulations that will become part of the fabric of each individual's life, it will be difficult to maintain illusions of freedom. In other words, we will experience creeping dictatorship, although perhaps no single figure will symbolize the centralization of authority. We will have industrial martial law.

Clearly, corporate growth in energy-consuming industries would stop or regress; corporations engaged in energy production would grow exuberantly. But the economics of energy rationing would mean government controls on all industries – prices, wages, and profits. We have experience with price and wage control, with tax restructuring, with money-flow control by the Federal Reserve Bank – but these are relatively minor economic control systems. We have no experience with the regulation of as pervasive an economic force as energy. We shall find that control of tax incentives, prices and wages, or money-flow will have no corrective effect whatever on the depression that an energy shortage will create. The brain trusts can't save us – when we're out of energy we're out of heartbeat, not just breath.

If the energy shortage of the 1970s and 1980s doesn't cripple the stock market by wiping out stock prices, then ZPG will. What is ZPG? Zero population growth is a widely advocated practice. Its aim is to head off the inevitable exhaustion of water, land, food, minerals, and energy that is the necessary consequence of an unchecked population explosion. On energy grounds alone, we should establish a zero population growth by the 1990s, certainly by the turn of the century.

The anticipation of two decades of energy shortages in America, followed by a tide of public opinion for ZPG during and following the energy-shortage years, at the least would plunge stock prices to all-time lows. A collapse of the stock market would be tragedy of the worst kind. Our entire financial world is linked to the stock market. It is linked to the phenomenon of growth – growth in corporate production. A collapse of stock prices spells the collapse of retirement funds, bank accounts, insurance pools, and so on.

If the market should survive the energy crisis because we solve our energy problems, can it survive ZPG?

It's high time our nation took a serious look at the marriage between energy and our economy.

A regulated way of life by 1984 is not at all the stuff of fiction. If even one-half of our projected energy gap of the 1980s should materialize, we would face a shortage of fifteen percent in our energy budget. That would mean a fifteen percent reduction in national capacity for the production of goods and services and a drastic reduction in employment.

Who will ration energy? What criteria will be used? How will the control of the production of goods and services affect our political and social institutions?

Already the "planners" are laying plans for the first phase of energy rationing. With the approach of the winter of 1971, gas distributors warned their customers to expect curtailment in the event of a severe winter. The New York Public Service Commission sent the following warning to gas customers:

> The first to be curtailed would be the factories and power plants that contract for gas on an interruptible or when-available basis and have the capacity to shift to alternate fuels: oil or low-sulphur coal.
>
> Next to be cut would be the industries with firm contracts that count on getting their full gas requirements. Commercial customers – shopping centers, department stores, theaters, restaurants – would be deprived only after the factories had been shut and as a next-to-last resort.
>
> And, finally, in the direst of crises, the human-needs customers – schools, public buildings, hospitals and homes – would go cold.[1]

The Supreme Court in a 7 to 0 decision has upheld the authority of the Federal Power Commission to allocate natural gas to homeowners rather than industrial users in the event of a shortage.[2] We can see in this decision the flow of economic power to a focal point in Washington and that the necessity for national survival will ultimately centralize energy rationing rights in the federal government.

Let's look at another example of how far thinking has progressed in the direction of "master planning" for energy rationing. In an interview reported in the *Los Angeles Times,* a vice president of Southern California Edison Company, which is currently experiencing severe gas shortages in conjunction with continued commercial and residential growth, made the following observations:

> It is our expectation that the consequences of an embargo on electric energy growth in terms of housing availability, construction activities, retail sales, employment and quality of life, especially for the disadvantaged, will create major repercussions when its import is felt by the general public.
>
> About one-third of Edison's growth load could be stopped if no further connection of new customers were allowed. To stop the remaining two-thirds of the utility's growth rate it may be necessary to terminate all appliance and equipment sales and generally establish an embargo on all retail activities involving electric energy using devices.[3]

England in 1972 provides an example of a devastating power shortage. The shortage was not brought about by loss of an energy source, such as we face in the coming decades. It resulted, rather, from a miners' strike. Nevertheless, the dark events during the six weeks of the strike are portents of what may well happen here over the next two decades.

The shortage of coal reduced electrical production. At the height of the strike, the total loss of electrical power was about fifteen percent. At the beginning of the strike, there were localized voltage reductions, or "brownouts." Then, certain heavy electrical users were forced to reduce their electrical consumption. Consequently, many industries were placed on a three-day work week. As the strike continued, entire residential areas and business districts were blacked out for as many as nine hours a day. Homes and businesses became dark and cold. Some industries were then forced to lay off substantial portions of their work force. However, layoffs in one industry affect job security in related industries. It makes little sense to make tires for cars if the cars are not coming off the assembly line. Layoffs became more widespread, affecting even industries that were otherwise not heavily dependent upon electrical power. By the sixth week of the strike, it was estimated that about three million Britons were unemployed. This figure constituted ten percent of the work force. The government was in a state of panic. It was feared that prolongation of the strike for but an additional two weeks would lead to layoffs of twenty million workers – two-thirds of the work force! In effect, England's industry would have completely shut down.

There is a breaking point. England was frighteningly close to it.

As we shall see, America faces an energy gap of about thirty percent in the 1980s. If a fifteen percent gap in electrical production alone brought England to the brink of disaster, how much more havoc would a prolonged and, possibly, *permanent* energy gap in total energy-availability wreak on the American economy, its institutions, its political stability, and the health of its people?

OUR NONRENEWABLE RESOURCES

Our energy crisis is "endemic and incurable" says John A. Carver, Jr., vice chairman of the Federal Power Commission.[4]

"We can anticipate that before the end of this century energy supplies will become so restricted as to halt economic developments around the world."[5]

"The furnaces of Pittsburgh are cold; the assembly lines of Detroit are still. In Los Angeles, a few gaunt survivors of a plague desperately till freeway center strips, backyards and outlying fields, hoping to raise a subsistence crop. London's offices are dark, its docks deserted."[6]

Let us briefly consider both the eternal and finite power sources. They represent our total power options.

Our dilemma is the extent of our dependence upon finite power sources. Industrial America is ninety-six percent powered by fossil fuels![7] We draw four percent of energy from nuclear and hydroelectric power. In other words, our present dependence upon fossil fuels is practically *absolute.*

Forgetting the enormous pollution problems posed by the burning of fossil fuels, we could probably use them forever *if* the earth's supply were unlimited. This is not the case. Fossil fuels are a nonrenewable source of energy. Once gone, they are gone forever.

Why can't we create fossil fuels?

Approximately 300 million years ago the geological processes that resulted in oil, gas, and coal formations were at their height. Nature still forms the fossil fuels but at a formation rate *one million times slower* than man's consumption rate. This imperceptably slow formation rate classifies fossil fuels as nonrenewable.

We consume in a year what nature took a million years to create. In the time it takes to read this sentence, we have consumed what nature took the better part of a year to create.

Nuclear fuels are also finite. There is only so much uranium or thorium or deuterium on earth; unlike the fossil fuels, no more can be created by man or by nature, even at a slow rate. The present reserves are our finite energy bank.

In distinction to finite power sources there are the eternal sources: sunshine, wind, tides, and flowing water. These sources will last as long as the sun continues to shine on the earth, offering man the possibility of converting these phenomena into power.

All the eternal power sources, solar power, wind power, tidal power, and hydroelectric power, if tapped, would yield their energy as electricity. The most powerful, but the most unobtainable of the eternal sources is sunshine; the most easily harnessed is flowing water (hydroelectric power).

Perhaps we shall someday be able to harness our potentially greatest source of energy. Unfortunately, as we shall see, the necessary solar technology appears to be relegated to the remote future.

Theoretically, in the United States, cars, homes, factories, and every need could be run by electric power. To do so would require about 2 trillion watts of power generation. Our hydroelectric power potential is only about five percent of this figure. Clearly, then, we have already expanded in numbers and living standard far beyond the capacity of our most accessible and renewable energy source to sustain us *even in the present.*

And we shall see that damming rivers for energy can lead to disastrous effects upon ecology, agriculture, local economies, and health.

Depending on the form in which it is used, energy manifests itself as electrical energy, mechanical energy, radiant energy, or heat energy. Power is the rate at which energy is used; it is a measure of energy expenditure per unit of time. Whereas we could choose from many units of measurement (B.T.U., calories, ergs, etc.) we'll measure energy (the capacity to do work) in terms of the joule, and power (the rate at which work is performed) in terms of the joule per second – which is a watt of power (1 watt of power equals 1 joule of energy per second). We will not be concerned with the joule and watt as such. We mention these terms only because they are used in the assessment of the lifespan of our resources.

What do the joule and watt really mean?

These measuring guides acquire meaning through application to situations with which we are all familiar.

For example, the average human operates at 100 watts (100 joules per second, 6,000 joules per minute), about equal to an average kitchen light bulb; his brain operates at 20 watts, and his body at 80 watts.

All industry, including transportation, in the United States is operating at two trillion watts; the world's at six trillion watts. Stated another way, the United States expends about one-third of all the industrial power on earth. Six percent of the earth's population uses thirty-three percent of its energy.

The present sources of United States power of about two trillion watts are approximately proportioned as follows:[8] coal, 25 percent; oil, 40 percent; gas, 31 percent; hydroelectric, 3 percent; nuclear, 1 percent.

Let us make a few further comparisons, approximations of muscle to industrial power. The United States industrial power expenditure is 10,000 watts per capita, or 100 times the biological power of a person. The world average is 1,500 watts per capita, or 15 times the biochemical power. In England the power expenditure per capita is 7,500 watts; in the U.S.S.R. it is 5,000; in Japan it is 3,000; and in India 500. These figures reflect relative living standards.[9]

However, we are not concerned with energy and power physically, but only in the relative sense of how long and how well our energy sources can serve us. By thus measuring energy and power we can assess the developing power shortage.

Whereas the concept of "lifespan" will be developed in subsequent chapters, the following lifespans of finite energy sources for the United States should provide a general idea of the difficulty in estimating the life of an energy source.

PROBABLE LIFESPAN OF
U.S. ENERGY RESOURCES[10]

Gas	40 years at the 1970 consumption rate, and less than 30 years at the present growth rate
Oil	20 years at the 1970 consumption rate, and less than 15 years at the present growth rate
Coal	200 to 300 years if coal is used to synthesize oil and gas at their present growth rates
Uranium	100 to 1,000 years after the breeder reactor is on-stream by the year 2000 or 2020 for a six-trillion-watt economy
Deuterium	over a billion years if we could develop controlled thermonuclear fusion reactors

Not apparent in this table are the grave problems of the *rate* of acquisition and deployment of energy from these and other sources. The employment of energy by United States industry is growing faster than the population. We have seen how this is an index of our rising standard of living, as more goods are created and more services are performed on a per capita basis.

As of 1970 the United States growth rate was 1.1 percent per year, while its energy growth rate was 7 percent per year.[11]

To maintain this seven percent growth rate requires an energy output that doubles in about ten years. With a doubling time of ten years, by 2000 we would be using *eight* times as much energy as in 1970!

Obviously, this compound rate can't be maintained indefinitely. The vital questions are: (1) when will population reach a zero growth rate? (2) when will our power production reach a steady-state, or zero growth? (3) when will we develop additional sources of energy? The future holds a trade-off – a massive power default or a massive population reduction, or both.

The following graph shows our present energy consumption trends and the various sources of energy.

UNITED STATES ENERGY EXPENDITURE[12]

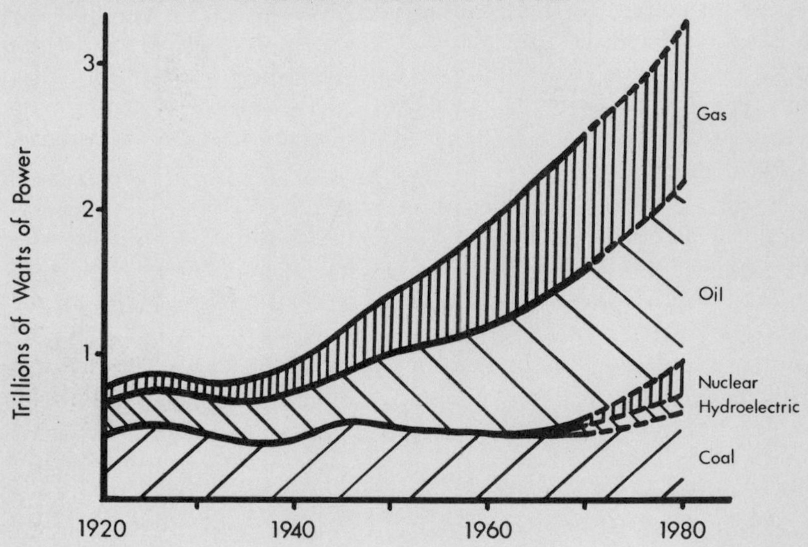

Further complicating the energy picture is the fact that population and energy consumption appear to be mutually stimulating. The figures show two "explosions" – the recent population explosion and the recent explosion of technological innovations.

Which caused which? The answer is highly debatable. However, it appears that the two growth trends stimulate each other: increases in population bear heavier on technology to sustain and exploit growth, and technological innovation permits an increase in population. Technological innovations, in turn, of course, are actuated by energy sources.

Just how advanced is the United States in the employment of energy compared to the rest of the world? If the population should hold constant around the globe and the rate of industrial growth were five percent per year for countries other than the United States, then the world would need 57 years to catch up to our present rate of energy employment, or our material standard of living. This figure is based on statistical analysis techniques such as those used to compute compound interest rates.

In fact, catch-up time may prove to be "leveling down" time. The fastest increasing populations are in underdeveloped countries, whereas the fastest industrially developing countries are the developed countries. The gap between the "haves" and the "have nots" is widening, not closing. The great leveler may turn out to be the power default. Even the "haves" must close up shop when the lights go out.

Indeed, "have-not" nations would probably fare better under a collapsing power base. Not as wedded to fossil fuel to begin with, they would be less affected by its withdrawal. An agrarian economy would still derive power from plows drawn by horses and oxen.

Suddenly denied the energy to run tractors and combines, how would modern industrialized society farm its land and feed its people?

Approximately one percent of all United States industrial energy is expressed on the farm.[13] One percent of our energy permits us to feed a hundred percent of our population! Without this energy we would clearly starve by the millions.

How so?

The energy output of all 200 million Americans as biochemical energy is equal to the energy now employed on the farm. This means that if *all* Americans were to work on the farm without the use of machinery, their "muscle power" would just equal the energy needed to produce what we now produce with the aid of machines.

"Back to nature," in this context, is a preposterous thought! Our industrial environment is so much more productive of the means of survival than methods of prehistory that there is no turning back to nature. If everyone were engaged in agriculture directly there would be no fertilizer industry or pesticide industry – both of which account for a fifty percent increase in acreage output. In our hypothetical example, farm output would drop fifty percent and would greatly affect the sustainable population. Back to nature or organic farming – any system without farm machinery, fertilizers, or pesticides – means death by starvation for 100 million Americans if land occupancy remained constant (it was fifty percent of all potentially arable land as of 1970),[14] or it means that 200 million Americans would have to develop and farm all the potentially arable land in the nation just for nutritional survival. Who would manufacture the necessities and the amenities of life?

Even so, were some of us to return to farming we would find that we have lost many traditional skills in moving from an agricultural to an industrial stance in the past fifty years. We have lost the farm animals necessary to till the land. Obviously, we could not bring them back overnight. They are of a past age, dead and gone except as nostalgic items in the memories of our senior citizens. Most important, we have lost the small farmer and his special blend of knowledge and skill. We have lost his love for the earth, his understanding of the land, and his ability to extract, through patient labor over long hours, those fruits that the earth so reluctantly yields. Even the gentle and industrious Amish would come to understand the extent to which they depend upon fossil fuels and the modern industrial complex to maintain their special life style.

It is ironic that we should be surpassed by the might of natural forces *operating through our inventions.* These inventions – the internal combustion engine and the electric motor, the telephone and the computer – have repositioned us into a new network of dependencies.

NOTES

1. "Gas Shortage Poses a Nationwide Threat of Cutbacks," *The New York Times,* November 21, 1971, p. 1.
2. *Chemical and Engineering News* (June 12, 1972), p. 5.
3. *Los Angeles Times,* January 23, 1972.
4. "Energy Crisis: Are We Running Out?" *Time* (June 12, 1972), p. 49.
5. John F. O'Leary, former director of the U.S. Bureau of Mines, in *Chemical and Engineering News* (January 3, 1972), p. 4.
6. *Time* (January 24, 1972), p. 32.
7. U.S. Bureau of Mines.
8. Ibid.
9. *Energy and Power, Scientific American* (September, 1971), p. 142.
10. Committee on Resources and Man, National Academy of Sciences, National Research Council, *Resources and Man* (San Francisco: W. H. Freeman and Co., 1969), and *Energy and Power, Scientific American* (September, 1971).
11. Committee on Resources and Man, *Resources and Man,* ch. 8.
12. U.S. Bureau of Mines.
13. Fred Singer, "Human Energy Production as a Process in the Biosphere," *Scientific American* (September, 1970), p. 175.
14. Committee on Resources and Man, *Resources and Man,* p. 67.

Oil: The Gathering International Dangers

As we have indicated in the chart on the probable lifespan of United States energy resources, our oil repositories are headed for exhaustion in two decades; hence, even if we solve our immediate oil shortage, the problem would soon reappear.

In the pages that follow we shall see that we are headed for oil rationing within a decade, and competition, possibly conflict, with our allies for access to oil from the Middle East.

In mid-1971, the Nixon Administration presented Congress with a plan for the long-range development of United States energy resources. This program relies heavily on nuclear power and, if implemented, will cost at least three billion dollars through 1980. Senator Henry M. Jackson (D.-Wash.), recognizing the importance of a comprehensive energy policy, expressed surprise that Nixon's message did not deal more with oil policy questions. Senator Jackson is among many political, industrial, military, and economic leaders who are fearful of the growing United States dependence on imports of foreign oil.

OUR OIL HERITAGE

How do the experts estimate the total amount of oil that exists in nature, hidden as it is among the vast rock strata of geological formations? By contrasting the oil discovered per foot of exploratory drilling with total feet of exploratory drilling for a given geological site, and then extrapolating to the total number of potential similar sites, one can get a good idea of the total amount of oil to be expected by drilling in all logical sites on earth. Naturally, these estimates are approximate.

In referring to a resource's heritage throughout this book, we are talking about the amount of that resource withdrawn plus the amount discovered (and not yet withdrawn from its repository) plus presumed amounts of that resource based on statistical analysis. Reference to reserves will always combine discovered and presumed resources. When discussing oil resources, we do not include nonliquid sources of oil, such as oil shale.

The total oil heritage of the United States (including Alaska) is about 200 billion barrels. Of this total, about fifty percent has been recovered and burned; about thirty percent has not yet been pumped from the oil fields; and about twenty percent is not yet discovered, but is presumed to exist, somewhere, on the scientific principles of geology.[1]

An analogous breakdown of oil for the world has also been made. Both are given in the table on page 16.

The amount of oil that is estimated to be ultimately recoverable in the United States varies widely from expert to expert, and ranges from 145 billion barrels to 590 billion barrels. However, most estimates based upon generally accepted statistical methods place our oil heritage at about 200 billion barrels. About 100 billion barrels of this has already been consumed. Hence, we shall use the figure of 100 billion barrels of oil as our reserve of oil that is recoverable under present technological practices.

Not only is it an art to estimate the total repository of any natural resource, but it is an art to project its ever-changing consumption rate, that is, its lifespan.

No natural resource is consumed at a constant rate, for obvious reasons. Consider the history of oil. When oil was first obtained from the ground, its rate of consumption was very low. Then came the diesel locomotive, the car, and the airplane, so that the rate of consumption rose exponentially. Now since this resource is finite, eventually the rate of consumption must decrease, also exponentially. Generally speaking, oil consumption exhibited an indistinct beginning during the 1880s, a prime usage during the 1900s, and, presumably, will show an indistinct phasing out in the twenty-first century. Obviously, the consumption rate is not constant, and since the beginning and end are so imprecisely flexible, one speaks of a lifespan in a very general sense. In fact, some experts prefer to use the "80 percent" lifespan, i.e., the period of time during which 80 percent of a resource is expected to be

OIL RESOURCES AND MAJOR REGIONS OF THE WORLD
(IN BILLIONS OF BARRELS[2])

	U.S.	+	U.S.S.R. & CHINA	+	FREE WORLD	=	WORLD
Total heritage	200		500		1,300		2,000
Consumed	100 (50%)		10		90		200 (10%)
Reserves	100 (30% discovered) (20% presumed)		490		1,210		1,800 (30% discovered) (60% presumed)

depleted. The first 10 percent takes too much tooling up and the last 10 percent is too difficult to extract.

We cannot argue the "last drop moment." In addition, as a resource becomes depleted, costs rise and substitutes are searched for before the resource is phased out gradually.

One way to conceive of the lifespan of oil is to make a calculation assuming that a constant rate of consumption will prevail. This is obviously never the case, but such estimates lend perspective to the impending power shortage.

For example, the estimated oil reserves of the United States are 100 billion barrels. The 1970 consumption rate was 5.4 billion barrels.[3] At this rate, our oil reserves would be depleted in about twenty years.

Actually, 1.2 billion barrels of this total were imported from Canada, Venezuela, and other countries; only 4.2 billion barrels were domestic.[4] However, should the United States be forced to rely upon its own oil reserves, we can see that these are perilously short-lived in terms of national self-sufficiency for our own lifetimes.

In order to place the variability of consumption rates in perspective, let us make a few constant-rate consumption calculations.

What world population could be supported at a constant rate of consumption equal to that in the United States (5.4 billion barrels per year) for oil reserves to last 10,000 years? Only five million, or one-fourth of New York State's population. On the other hand, if the 3.5 billion people in the world enjoyed a United States living standard (and oil consumption rate), world oil reserves would last ten to twenty years. Here we see that the high living standard aspirations of man cannot be fulfilled as long as we are dependent upon oil as the main power resource.

The previous calculations assume that oil will be sold and used freely among the nations of the free world on a business-as-usual basis.

Suppose the experts have underestimated our oil reserves. Let us say that the oil reserves of the United States prove to be twice as large as the present estimates. Are we saved? Our oil allowance would last not for two or three but for four or six decades, which is also a very short lifespan. But even this is optimistic. Consumption of energy appears to follow Parkinson's law: the amount of energy consumed rises to meet the amount of energy *available.*

WORLD OIL RESERVES AND THE MIDDLE EAST

Oil is not evenly distributed by geographic region. Below is a regional distribution of world oil reserves. Note the Middle East figure. It represents about sixty percent of all free world reserves, exclusive of United States reserves.

APPROXIMATE WORLD OIL
RESERVES BY GEOGRAPHIC REGION[5]

REGION	BILLIONS OF BARRELS	PERCENT OF TOTAL
U.S.	100	6
Western Europe	20	1
Africa	170	10
South America	190	11
Middle East	740	40
Other Regions	90	5
U.S.S.R. & China	490	27
	1,800	100

Let's consider the oil self-sufficiency of several regions and the oil dependency of one upon another. When this is done, the precarious power position of the United States and our allies becomes more apparent.

We'll examine three major oil-consuming regions: the United States, Europe, and Japan. First, let us look at the approximate oil reserves of each region and its probable lifespan of domestic reserves should each be forced to become self-sufficient, while trying to maintain its present rate of oil consumption.

LIFESPAN OF OIL IF EACH REGION WERE FORCED TO RELY ON ITS OWN RESOURCES (BILLIONS OF BARRELS)[6]

REGION	RESERVES	ANNUAL CONSUMPTION	ANNUAL GROWTH RATE IN CONSUMPTION	APPROXIMATE LIFESPAN (CONSTANT RATE)
United States	100	5.4	5%	20 years
Western Europe	20	4.7	10%	5 years
Japan	negligible	1.7	17%	–

The United States oil imports of about 22 percent break down as follows: 10 percent from Venezuela, 4 percent from Canada, 4 percent from the Middle East and Libya, and 4 percent from other sources.

By 1985, we shall probably be using in excess of 10 billion barrels of oil per year (contrasted with 5.4 billion in 1970), but obtaining approximately 60 percent (6 billion barrels) from foreign sources; 46 percent of our oil will come from Africa (mostly Libya) and the Middle East.[7] At the same time, Japan and Western Europe will be making similar demands from the oil-rich countries of Africa and the Middle East.

Added to this projection of oil consumption and oil procurement needs is another dimension – a political and economic force.

The twelve-year-old Organization of Petroleum Exporting Countries (OPEC) has decided that, upon expiration in 1976 of its present pricing contracts with oil-consuming countries, it will raise the price of oil and demand the right to buy at least 20 percent in the oil companies that develop its oil fields.[8] Eventually, the OPEC countries, which consist of Algeria, Nigeria, Venezuela, Iran, Iraq, Qatar, Saudi Arabia, Libya, Indonesia, Abu Dhabi, and Kuwait, plan to own 51 percent of foreign oil companies.[9]

Indeed, the Arabian American Oil Co. (Aramco) has already agreed to sell 20 percent of its ownership to Saudi Arabia. Aramco is the first consortium of oil companies (Standard Oil of California, Jersey Standard, Texaco, and Mobil) to so yield to OPEC. OPEC has established a "bail-out" fund just in case any one of its member nations is boycotted by the purchasing companies in retaliation.

Clearly, the Third World means business.

The OPEC countries presently produce 90 percent of the world's oil exports.[10] They plan to train their own technicians to explore and develop oil fields as well as to develop their own oil refining capacity. All this is a prelude to the 51 percent takeover, starting with the expiration of present oil field leases beginning in 1983.

Oil price increases are inevitable in 1976. Such an increase would be especially adverse for Western Europe and Japan, which import about ninety percent of their oil from the Middle East and Africa; and initially less adverse for the United States, where oil imports from this region may be only six to ten percent of our oil budget instead of the present four percent. But by the 1980s, we have seen the figure will be about forty-six percent. In fact, our oil companies could be owned to the same degree by foreign interests.

In addition to high consumption rates, the previous chart shows that the United States, Western Europe, and Japan are experiencing very rapid growth rates in oil consumption. Obviously, this growth cannot be sustained indefinitely.

We have seen from the world oil chart that the Middle East alone possesses 740 billion barrels of discovered plus presumed oil reserves. Of these, less than half have been proved. At a constant consumption rate, if all *proved* Middle East oil went to Western Europe and Japan exclusively (combined consumption of 6.4 billion barrels), this oil would be depleted in about forty-five years. Actually, the consumption rate of oil in Japan and Western Europe is doubling every six years – which gives us their charted weighted averages of seventeen and ten percent growth rate. If it were possible to maintain these growth rates, the *proved* Middle East and African oil would be depleted in less than twelve years by Japan and Western Europe alone.

Actually, it is not possible to maintain this growth rate for a number of reasons. The maximum oil that can be obtained from an oil field is related to the rate of oil withdrawal. An excessively fast rate is not desirable. Three of the largest oil-producing countries, Libya, Venezuela, and Kuwait, have already limited their daily production quotas.

If, as a result of political, economic, or military activities, the United States, Japan, and Western Europe were forced to rely on their combined reserves of 120 billion barrels, then, at their present combined annual consumption rate of 11.8 billion barrels, our oil would be depleted in about 10 years. This figure would fall to less than 7 years if we took into account oil consumption growth rates for these regions.

Obviously, if Japan, Western Europe, and the United States were totally shut off from Middle East and African oil, rationing would be necessary immediately.

Even without a political crisis, the lifespan of free world oil is relatively short. All the nations of the free world (all non-Communist bloc countries) consume 14.6 billion barrels of oil per year at present. With a total free world reserve of 550 billion barrels of *proved* reserves, that indicates a lifespan of approximately 35 years.[11]

It is estimated in the world oil resources chart that the oil that will one day be discovered is more than twice the presently proved reserves. However, the accelerating rate of oil consumption per year will reduce any projected lifespan of these new discoveries.

Hence an estimate of three to four decades as the lifespan of free world oil is probably the maximum upper limit. It is probably much shorter.

IS THE ANSWER IN ALASKA OR IN NUCLEAR "STIMULATION"?

Is Alaska the answer to our oil needs? Not in a long-range sense. The Alaskan finds represent about 20 percent, or 20 billion barrels of oil, of all domestic oil reserves – at a reasonable estimation.[12] Even if all the oil proves to be there, and recoverable, it could supply our needs for only about four years, based on our current 5.4 billion barrels of oil consumption annually. This is not to imply that this oil can be withdrawn, shipped, and processed in four years. The rate of withdrawal will be spread over many years.

At the most optimistic estimates of recoverable oil (about 30 to 50 billion barrels),[13] it would last six to ten years.

Even so, Alaskan oil has created a controversy among ecologists, who fear that an oil spill would damage the Arctic tundra. Hence work on the Alaskan pipeline is being delayed.

An oil spill could damage 0.01 percent of the Arctic tundra and Alaskan forests combined.[14] (Most people think that an oil spill would cover Alaska.) We should be concerned, but we should also place it in perspective to United States energy security. At a time when we are importing 22 percent of our oil needs and moving toward 60 percent in 1985, we need to order priorities. Let's take every precaution to spare our environment, but we desperately need the oil – now.

Is nuclear stimulation the answer to our oil needs? Dr. Seaborg has raised the question, proposed by many scientists, of using atomic explosives to win oil from oil shale in his recent book, *Man and Atom.*

Any such scheme to obtain *meaningful quantities* of oil can only fail – and could be dangerous (as in the example of gas stimulation to be discussed later).

"Oil shale" does not contain oil but hydrocarbon solids, the molecular constituents of oil in solid form. Shale must be mined, like coal, not drilled like our oil and gas deposits. The shales hold about 25 gallons, or less, of potential oil per ton of shale.[15] A nuclear bomb could be exploded underground to generate a cavity in which the oil could be distilled and pumped out.

All our oil shale strata probably contain hundreds of times the hydrocarbon content of all oil deposits. Hence the interest in shale deposits. But is it recoverable by nuclear blasting?

A 50-kiloton atomic bomb (equivalent to 50,000 tons of TNT, but possessing approximately 3 kilograms, or 6 pounds, of fissionable uranium) would pulverize about one million tons of rock. It would be detonated at a depth of thousands of feet in oil shale strata. This would yield about half a million barrels of oil, at best.[16]

Recall that our 1970 consumption rate of oil was about 5.4 billion barrels. Could we obtain this oil from shale by nuclear stimulation? At the rate of 0.5 million barrels per 50-kiloton atomic bomb, *we need 10,000 such blasts to yield 5 billion barrels of oil.*

Can 10,000 nuclear shots per year (twenty-seven a day!) be safe and economical?

What about nuclear stimulation of oil wells? Only one-half of the oil in oil fields can be pumped out by ordinary means.[17] There are hundreds of thousands of oil wells in our country (600,000 producing wells in the world). If we stimulate ten percent of them by atomic explosions, then we are back again to thousands of underground nuclear explosions per year.

If nuclear stimulation is impractical, is it not possible to obtain oil from oil shale by conventional mining techniques? Unfortunately, the energy required for total processing is greater than the energy obtained in the oil for about 99.99 percent of all oil shales.[18] Even though there is "oil" in oil shale, most of it is too dispersed to obtain with a net energy gain.

What shall we do as the depletion of oil approaches? Ration oil? Suffer a lower living standard? Synthesize "oil" from coal? Import oil and hence become progressively more dependent upon foreign sources for our "lifeblood"? These questions will be studied in the next several chapters. However, there appears to be no answer to the oil shortage. There is no way to wrest centuries of oil supply from nature; she simply has not much more to give than about two decades' worth of domestic oil.

The oil companies appear to be fully aware of the inevitable decline in their power and have been doing something about it. At least eighteen oil companies have already invested in the production and processing of uranium. As of 1969, oil companies conducted approximately forty percent of exploratory drillings for uranium.[19]

Oil companies currently own approximately forty-five percent of known uranium reserves, and approximately thirty percent of known coal reserves.[20] If they were not aware of the impending energy crisis, why would they be investing so heavily in alternative energy sources?

NOTES

1. Committee on Resources and Man, National Academy of Sciences, National Research Council, *Resources and Man* (San Francisco: W. H. Freeman and Co., 1969), ch. 8.
2. There is considerable variation in published figures for oil reserves. These figures are adapted from *Resources and Man,* table 8.2, p. 194, and represent an optimistic estimate of total world reserves of oil.
3. "U.S. Energy Outlook, An Interim Report of the National Petroleum Council," vol. 1, July 1971.
4. Ibid.
5. Adapted from: Committee on Resources and Man, *Resources and Man,* ch. 8.
6. Adapted from: *Energy and Power, Scientific American* (September, 1971), pp. 134-144.
7. "U.S. Energy Outlook," vol. 1, July 1971.
8. "Oil: Nationalization in Part," *Time* (March 27, 1972), p. 88.
9. "Facing a Powerful Cartel," *Time* (January 24, 1972), p. 59.
10. "The Arabs and Their Oil," *Chemical and Engineering News* (November 16, 1970), pp. 58-71.
11. Ibid.

12. "U.S. Energy Outlook," vol. 1, July 1971.
13. *Energy and Power, Scientific American* (September 1971).
14. Based on data in *Chemical and Engineering News* (April 3, 1972).
15. Glenn T. Seaborg and William R. Corliss, *Man and Atom* (New York: E. P. Dutton and Co., 1971).
16. Bernard L. Cohen, *The Heart of the Atom* (Garden City, N.Y.: Doubleday-Anchor Books, 1967), p. 83.
17. Seaborg and Corliss, *Man and Atom,* p. 180.
18. Committee on Resources and Man, *Resources and Man,* ch. 8.
19. Neil Fabricant and Robert M. Hallman, *Toward a Rational Power Policy* (New York: George Brazillier, 1971), p. 97.
20. Ibid.

The Futility of Our Natural Gas Situation and Strategy

We face an immediate natural gas shortage of great proportions, protracted over the next two decades, with no hope of obtaining natural gas by "stimulating" existing gas fields.

Either we will have to import natural gas, as many experts suggest, or curtail consumption. Importing gas will make America vulnerable to foreign influences; curtailing gas consumption will thwart industrial growth and living-standard levels.

As long ago as November 23, 1971, *The New York Times* reported that natural gas distributors around the country were alerting their customers to possible curtailments in the event of a cold winter. Some gas companies were refusing to accept new large customers, and even some residences. In the event of actual cutoffs, New York State would be among the first, probably starting with the winter of 1972-73. The problems were attributed by the natural gas industry to the fact that consumption is outpacing the discovery of new deposits at an ominous rate.

We have not fully developed our gas fields to provide for the great quantities of gas anticipated by industry. The interim gas is there; we have failed to explore and develop our resources in time to meet the needs of the 1970s. But the overall gas problem is another matter – we will be almost at the end of natural gas before the year 2000. There just isn't enough in nature's repositories to keep up with our consumption.

OUR NATURAL GAS HERITAGE

The occurrence of natural gas deposits closely parallels oil deposits. The discovery ratio of oil to natural gas is almost a constant 6,000 cubic feet per barrel of oil.[1]

The total energy potential of all natural gas is roughly 1.6 times as great as that of oil. From the total estimated world resources of gas and from consumption-rate trends, we can estimate a lifespan for gas: about two to three decades for the United States, similar to that for oil.

Estimates of the amounts of gas that will be ultimately recovered in the United States vary widely from expert to expert. Below are a few estimates from the more prestigious sources to indicate the range of values for natural gas.

U.S. HERITAGE OF GAS IN TRILLIONS OF CUBIC FEET

National Academy of Sciences	1,000
Potential Gas Committee (Organization of Commercial Gas Companies)	1,290
Hendricks[2]	2,000

We have chosen an average estimate of 1,400 trillion cubic feet as our original heritage of gas. About 400 trilllion cubic feet have already been consumed. This leaves us with about 1,000 trillion cubic feet of gas as our present reserve. This figure conforms to the most frequently quoted figure in the literature, and it will be used as an approximation of our gas reserves that are recoverable, using present technological practices. Of this 1,000 trillion cubic feet of reserves, about three-fourths are yet to be discovered.[3]

As with oil, we face rising gas imports and ever-growing dependency on foreign sources. This dangerous trend toward loss of energy self-sufficiency is evident in the National Petroleum Council figures: In 1970, the United States imported 22 percent of its oil and 4 percent of its gas. The Council projects that by 1985 we will be importing 60 percent of our oil and 28 percent of our gas.

On a total energy basis (fossil fuels, nuclear, and hydroelectric), the United States was about 90 percent self-sufficient in 1970, but by 1985 this will deteriorate to drastically reduced percentages.[4]

By 1975 our gas gap will widen to about 5 trillion cubic feet. Five trillion cubic feet of gas represents a shortage of about 20 percent; by 1985 this shortage could rise to over 13 trillion cubic feet (28 percent) of gas needs.[5]

UNITED STATES GAS DEMANDS EXCEED SUPPLY[6]

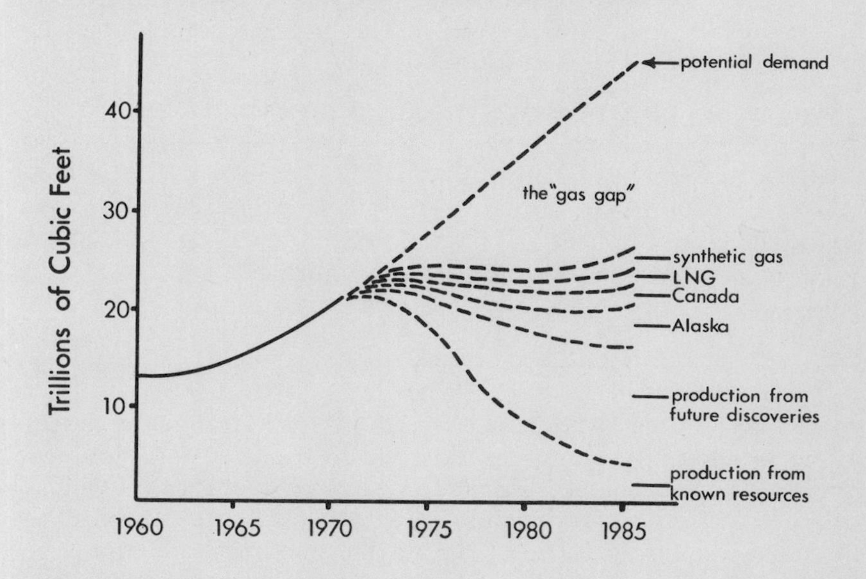

We can see from the chart that we can expect only small percentages of our gas needs from liquefied natural gas imported from abroad, mostly from Algeria and Russia. President Houari Boumédienne of Algeria has warned: "Until now, the riches of the third world have served the interests of the rich nations. It is time for those nations to understand that economic colonialism – like political colonialism before it – must vanish."[7] And we can expect only small gas percentages derived from the oil fields of Alaska or Canada. Mr. J. J. Greene, head of Canada's Energy Ministry, warned the United States, "It would be wrong for your industry and your policymakers to look to Canadian supplies [of gas] as a panacea."[8] The production of synthetic gas from coal will be covered in the chapter on our synthetic fuel strategy.

Disregarding the implications of our turning to foreign suppliers, the then Secretary of Commerce, Maurice Stans, after a trip to Russia, spoke hopefully of United States gas companies contracting for gas with the Soviet Union.

Gas could be imported from Russia by first liquefying it at −260° Fahrenheit and then transporting it as a liquid in supertankers equipped with special "thermos bottle" compartments. Each cubic foot of liquid will generate 600 cubic feet of gas after it is allowed to reach normal temperatures.

Some unresolved questions are who will finance the plants to liquefy the gas for shipment to the United States, or tanker construction (3 billion dollars for 40 tankers), or the cost of the gas, estimated at about 1 billion dollars annually for a mere 4 percent of our gas needs. As the Middle East is to oil, so Russia is to natural gas; hers are the world's largest resources, as we will see in a later chapter.

NATURAL GAS AS THE SAVIOUR OF OIL?

The energy of natural gas (the energy released upon combustion) is about 1.1 million joules per cubic foot. The energy equivalent of oil is about 6 billion joules per barrel. If we multiply these energy equivalencies by the amounts of gas and oil available in the United States for possible future consumption – 1,000 trillion cubic feet of gas and 100 billion barrels of oil – we find, as was already indicated, that the energy banks of gas and oil are grossly equal in magnitude – 1.6 to 1 of gas to oil. Hence, gas cannot substitute for oil as an energy source. They are both in short supply.

Only coal, the giant of the fossil fuels, can be considered a substitute for oil or gas. This is the subject of the next chapter.

Natural gas, primarily methane, has two major uses: heating and electric power production. A possible third use is in a natural-gas-powered automobile. The gas would be contained in liquid form and burned in an internal combustion-style engine. The advantage of natural gas over gasoline is that natural gas, which is low in sulphur, will burn cleaner and will emit far less pollutants per mile than will gasoline.

However, converting the internal combustion engine to accommodate natural gas is not a long-range answer to the impending oil shortage. Since the energy equivalent of discovered and presumed reserves of gas is only equal to that for oil, we would be obtaining only a few decades of life for this internal combustion phenomenon for motive power.

The eventual demise of the internal combustion engine for transportation would be deferred only briefly by substituting natural gas for gasoline.

Natural gas is also being used in place of oil to generate electricity through the steam-driven turbine, which should further shorten the lifespan of natural gas.

PROJECT GAS-BUGGY

Some scientists have suggested using nuclear bombs to stimulate natural gas deposits to yield more gas, either for wells that are running dry or for deposits locked in impervious rock strata.[9] This suggestion qualitatively is good. It could work, and would yield more natural gas. However, for our gas needs, the quantities would be altogether inadequate. Indeed, its acquisition could be dangerous, even for small portions of our gas needs. For example, let's examine Project Gas-buggy – a cooperative venture between the El Paso Natural Gas Company and the Plowshare Program of the Atomic Energy Commission. The Plowshare Program is a research endeavor to use nuclear explosions for planetary engineering purposes, such as excavating harbors, water reservoir basins, and canals.

To stimulate natural gas emission from the Pictured Cliffs Formation, a rock strata near Farmington, New Mexico, on January 10, 1968, a 29-kiloton atomic bomb was exploded at a depth of about 4,000 feet. The blast created a crushed rock cavity of about one-half million cubic feet. Between then and November, 1969, some 300 million cubic feet of gas was extracted. Almost all radioactivity was contained in the cavity. The blast produced no radiation problems of any significant magnitude and Project Gas-buggy was a huge success. But the safety of an experimental blast is not in question.

In approximately one year, this experiment yielded 300 million cubic feet of gas. What volume it will ultimately yield no one knows. The purpose of the experiment was to provide such information. But, assuming ten years of such good gas flow (a reasonable estimate for the lifespan of a gas well), one could expect 3,000 million cubic feet, a very optimistic estimate. It is also a drop in the bucket. It is a woefully inadequate return for the waste of technical manpower and uranium resources.

Our 1970 annual gas consumption rate of 23 trillion cubic feet equals about 7,000 times the ultimate yield of one 29-kiloton bomb. At this "stimulation" rate we would need *7,000 bombs per year* to keep up.

To stimulate our gas fields for 10 years would require 19 atomic explosions per day! This is obviously a preposterous way to close the gas gap.

This scheme would entail the construction of more nuclear bombs than we presently have in our entire nuclear arsenal and their replacement every year for the lifespan of oil and gas – two to three decades – to achieve an increase of lifespan of some 20 to 30 percent!

Figure in this calculation the deployment of scientific talent that could otherwise be solving the problems of fusion power and the electrical problems intrinsic to our electronic future. How wasteful such schemes really would be of scientists, engineers, and uranium resources.

The total energy of uranium expended in nuclear stimulation for oil and gas would be about ten percent of the energy content of the obtained oil and gas. This is not a very great energy gain considering the limited resources of uranium in the nation.

Where, then, if we continue to consume more and more natural gas over the next decade, are future gas supplies to come from?

NOTES

1. Committee on Resources and Man, National Academy of Sciences, National Research Council, *Resources and Man* (San Francisco: W. H. Freeman and Co., 1969), p. 187.
2. T. A. Hendricks, "Resources of Oil, Gas, and Natural-Gas Liquids in the United States and the World," U.S. Geological Survey Circular 522 (Washington, D.C.: Government Printing Office, 1965).
3. Federal Power Commission, staff report, 1971.
4. "Energy Crisis: Are We Running Out?" *Time* (June 12, 1972).
5. "Gas from Coal, Fuel of the Future," *Environmental Science and Technology* 5 (December, 1971), pp. 1178-1182.
6. Sources: Humble Oil and Refining Company and the Federal Power Commission Staff Report.
7. "The Algerians Intend to Go It Alone, Raise Hell, Hold Out and Grow," *New York Times* Magazine, April 23, 1972, p. 18.
8. Ralph E. Lapp, "We're Running Out of Gas," *New York Times*, March 19, 1972, p. 42.
9. Glenn T. Seaborg and William R. Corliss, *Man and Atom* (New York: E. P. Dutton and Co., 1971), pp. 177-180.

Coal:
The Source of Synthetic Fuels

Our coal energy bank is theoretically large enough for at least a thousand years, assuming the present rate of consumption will remain constant. Coal is a possible solution to our impending gas and oil shortage: that is, synthesize them from coal. This could shorten coal's lifespan to about two centuries.

However, our present plans to produce synthetic fuels (gas and oil) from coal are much too slow to avert the onrushing shortage and consequent dependency upon foreign sources. Also, the ramifications of our synthetic fuel strategy could be dangerous to our environmental well-being and our long-range needs of coal as a chemical resource.

Coal is the giant of the fossil fuels, accounting for ninety-six percent of available energy from the fossil sources. But it inflicts a terrible penalty upon those who would wrest it from its sanctuary within the earth. Unlike oil and natural gas, which are pumped from beneath the surface, coal must be mined.

Until relatively recently, most coal was obtained from mines dug deep into the earth. Deep coal mining is dangerous to life, hazardous to health, and expensive. In a coal mine tragedy in South Africa in 1972, 450 miners perished in an explosion of mine gas, and many coal miners suffer from "black lung" disease, caused by the year-round inhalation of coal dust. Coal's cost in wages alone runs to about $2.75 per ton. On the positive side, the landscape above the mines remains relatively unscathed.

However, to mine all the coal that is presently taken by strip-mining would require an additional 78,000 coal miners, $3.5 billion of capital outlay, and about 5 years to develop the capacity to fulfill strip-mining productivity (about 300 million tons per year).[1]

STRIP-MINING AS A FUTURE

Deep coal mining had to give way to the faster and cheaper method (50 cents a ton in wages) of strip-mining. James Branscome, a native of Appalachia and director of Save Our Kentucky, Inc., made the point:[2]

> What made a [property owner in Kentucky] angry is a relatively new, cheap and easy method of mining coal. It is cheap and easy for the coal operators but expensive and difficult for the people who live in Appalachia. Because it is cheap and easy in the short run, strip-mining this year is expected to account for half of the coal produced in America. The coal will fuel power plants that light the sprawling suburbs and the dying cores of American cities. As a result, the hills and mountains of Appalachia will be irreparably scarred. The Appalachian people will be forced to flee their homes because of landslides and flooding. The entire region will look more and more like the flayed back of a man, the lifeless or heavily damaged pulp of a miscreant who sinned against industrial America.

How is strip-mining accomplished? After a seam is discovered, bulldozers and land movers cut a huge swath through the timber to get to the seam.

The seam itself is covered by a network of trees, rocks, soil, and plants. These are referred to as the "overburden." It is blasted loose and bulldozed into huge piles, or "spoil banks." Without the network of vegetation and their associated deep root systems, the spoil bank begins to erode. It may move slowly, inch by inch, like a great gray glacier, inexorably destroying everything in its path. Or it may erupt in the form of a sudden, catastrophic avalanche, burying homes and farms under its rubble.

When the overburden has been removed, giant augurs bore hundreds of feet into the seam and spiral out the coal. Alternatively, a huge monster-type steam shovel, ten stories high and as wide as an eight-lane highway, plunges giant jaws into the earth and scoops out 6,000 cubic feet with each bite. All the land around is utterly ravaged and it may be years before it will again yield vegetation.

The tragedy does not end here, as James Branscome points out:

> Appalachia receives heavy rainfall throughout the year, averaging around 45 inches. When it rains, the 20,000 miles of strip-mine benches in nine mountain states become chemical factories. The exposed rock and soil are rich in iron, manganese and sulfates, which combine readily with water to form corrosive compounds and acids that sterilize streams and poison wells.[3]

The greatest tragedy of all is that strip-mining appears to be our only hope of temporarily closing our energy gap, of buying time to permit us to phase in nuclear energy.

Coal can be burned to yield electricity. But it is dirty and pollutes the air. About one-half of our present electricity is obtained from coal.[4] It cannot be burned in cars, but it can be forced to yield synthetic gas and oil. What is more, the production of coal must be increased a thousand percent over the next ten years. The nine states making up Appalachia will not be the only ones to fall under the maw of gaint strip-mining machinery. The coal fields west of the Mississippi hold almost eighty percent of the coal that may be obtained by strip-mining techniques.[5]

Our total estimated coal reserve is 1,500 billion (1.5 trillion) tons. This includes coal discovered and not yet mined plus that presumed to exist. The total coal recoverable by strip-mining is

128 billion tons, or about 8 percent of our total coal reserves.[6] Since strip-mining is cheap and yields coal quickly, it is the route we shall be compelled to take. But these coal fields will be exhausted within 2 decades. Strip-mining will leave in its wake 71,000 square miles of desecrated land.[7] This is equal to the total land area of Pennsylvania and West Virginia. Coal obtained through strip-mining is the bitter harvest of our years of energy neglect.

WORLD COAL HERITAGE

How do we estimate earth's coal heritage? Several factors make this task difficult: (1) About fifty percent of the total coal underground is technologically nonrecoverable; (2) the degree to which we can improve this recovery rate is not yet known; and (3) recovery of coal is given in terms of a percentage of seams of a given width and depth in ground. These factors have been considered in the subsequent estimates for coal reserves.

WORLD RESOURCES OF COAL[8]

REGION	PERCENT OF TOTAL ESTIMATED COAL (7.6 trillion tons)
U.S.S.R.	56.5
United States	19.5
Asia	9.5
Canada	7.8
Europe	5.0
Africa	1.3
South America	0.3

To assess the lifespan of coal is another matter. One must make assumptions about: population growth trends; the proportion of energy that will be derived from coal for heating and the generation of electricity; the quantities of coal needed to make coke, which is essential in smelting iron ore for the production of steel (a chemical process that not even nuclear energy can duplicate); the use of coal as a possible raw material out of which to create a synthetic oil and gas as a replacement for oil; the use of coal for bacterial generation of proteins; and so on.

Also, the degree to which we will not be dependent upon coal for energy due to the development of hydroelectric and nuclear power is yet another factor in assessing the lifespan of coal.

Given below are two estimates for a world lifespan of coal, assuming a constant rate of consumption at different points in history. Estimates are based on a comparison of power output of the world economy with the energy bank of coal. As with oil and natural gas, a constant rate of consumption is unlikely. Nevertheless the calculations proportion coal's lifespan into some rough future perspective. Whatever the actual consumption rate of coal proves to be, we can see that the total energy potential of the world's coal will certainly be claimed by man in a scant several centuries. Considering that mankind is about 1 to 2 million years old as a species, a few hundred years of support from our fossil fuels seems short indeed.

•Based on oil and gas reserves being near exhaustion by the year 2020, and the world drawing its energy solely from coal, coal would have a "constant-consumption lifespan" of about three to four centuries.

•At the 1970 constant rate of world coal consumption, coal would last for 1,700 years.

Let's put coal's lifespan in further perspective:

1,700 years: coal's lifespan if only coal were used for energy in the world and the present rate of energy expenditure were held constant, i.e., there were no increase in population and no improvement in the living standards of the earth's population.

400 years: coal's lifespan if the population of the world doubled and energy consumption per capita were to double and then hold constant. The energy expended would be about four times the present world rate.

300 years: coal's lifespan if only coal were used for energy in the world and the average expenditure of energy in other nations were equal to that in the United States today, and were to hold constant.

UNITED STATES COAL HERITAGE

Upon combustion, our fossil fuel reserves will release the following proportions of energy: oil, 2 percent; gas, 2 percent; coal, 96 percent.[9] How long will United States coal reserves, 1.5 trillion tons, last? Our 1970 consumption rate was 600 million tons. As with the previous examples for world coal resources, the answer depends upon how fast coal is burned. An indication of how variable the lifespan of coal can be is seen in the table below.[10] It shows how long United States coal would last if burned to supply all the industrial energy for the year given.

COAL WOULD LAST (IN YEARS):	AT ENERGY NEED IN:
1,000,000	1776
30,000	1860
700	1970
200	2000

If our 1.5 trillion tons were to go to any exclusive use, it would break down as follows, based on constant 1970 United States consumption rates. These figures are based upon the hydrocarbon content of coal compared to the conversion efficiency to oil, the bacterial conversion efficiency to protein, and the chemical conversion efficiency by weight to plastics.

for synthetic oil:	250 years
for energy:	700 years
for proteins:	100,000 years
for plastics:	150,000 years

But coal can't do all of these things simultaneously. Coal is more long-lived as a source of chemicals than as a source of energy.

SYNTHETIC OIL PROSPECTS

Let us assume that the conversion of coal into a synthetic oil was chosen as the strategy by which to solve our oil problem. What would our energy situation look like under this condition?

Let us also assume that the rate of present oil consumption were to hold constant; that we would arbitrarily assign half our coal for iron smelting, electrical generation, and heating purposes; and that the conversion efficiency of coal into oil is about 20 percent. The conversion of coal into gas is about 30 percent efficient;[11] 70 percent of the energy of the coal is lost in the conversion process. The conversion of coal into oil (or gasoline) is less efficient than that for synthetic gas.

On the basis of the above, the United States could produce synthetic oil from coal, as Germany did during World War II, and maintain a 150-year supply. Europe could produce a 50-year supply.

The U.S.S.R., with greater reserves of coal, and a lower oil consumption rate, could provide domestic synthetic oil for 800 years.

The U.S.S.R.'s capability of producing synthetic oil for approximately eight centuries reflects not only their greater coal reserves but a lower oil consumption rate. If their standard of living were to approach ours, the life expectancy of synthetic oil from their coal would be shorter.

Clearly, these are gross estimations of the quantities of synthetic oil that could be produced from coal. The production of synthetic oil is a very complicated chemical process. In the briefest of terms, synthetic oil is produced by passing powdered coal, steam, and air over catalysts. The conditions of temperature, pressure, and recycling paths determine the nature of the final product, which could be kerosene for jet fuel, gasoline for a car, oil for a power plant, and many other products.

However, the rate at which the United States is planning for the research and development of synthetic fuel processes is tragically slow. By 1985, we expect to have only about 10 liquid fuel plants each of 100,000 barrels per day capacity. This represents about 365 million barrels per year of synthetic oil. Based on our 5.4 billion barrel consumption rate (double by 1985), that's roughly 3.5 percent of our projected annual needs in 1985.[12]

SYNTHETIC GAS PROSPECTS

Congress, in conjunction with industrial groups, is planning to launch the nation on a coal gasification program.[13]

Plants to convert coal into gas (coal gasification) are now in the experimental stages. Whereas the chemical process is known, the details of massive plant design must be worked out. By 1976, FMC Corporation will have one plant in operation that should produce about 90 billion cubic feet of gas per year from 10 million tons of coal per year.[14] [FMC is one of many corporations that contract with the federal government through the Office of Coal Research (OCR) for funds with which to launch a study of the many chemical processes that can convert coal into gas.] There are four major chemical transformation routes being studied – this gives one a sense of the complexity of the synthetic fuel technology.

SYNTHETIC FUEL PRODUCTION[15]

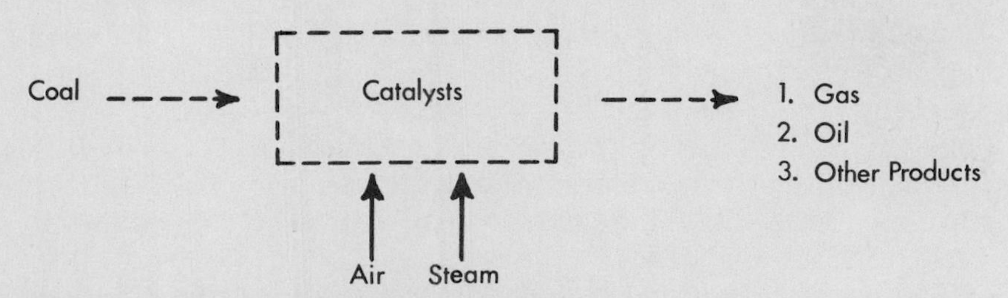

Operating conditions for coal gasification:
(1) 10 million tons of coal per year yielding 90 billion cubic feet of gas per year is an energy loss of 67 percent!
(2) a 90-billion-cubic-feet-per-year plant investment is $200 million or $50 billion for 250 plants to meet our 1970 gas needs of 23 trillion cubic feet.

The National Petroleum Council estimates that by 1985 we could have about 26 such plants in operation. Will these plants solve our gas shortage? Not at all.

By 1985 we can expect *no more than five percent* of our annual gas needs to be filled by synthetic gas.

At present gas needs of 23 trillion cubic feet per year (double by 1985), by 1985 those 26 plants would be producing a combined total of 2.3 trillion cubic feet per year of gas. A gas disaster looms.

No doubt synthetic fuel plant programs could be accelerated, but by how much? Present estimates indicate that a capital outlay of about $25 billion would be needed by 1985 in order to synthesize enough gas to equal projected imports (13 trillion cubic feet) to close our gas gap.[16]

Ralph Lapp (in *The New York Times,* March 19, 1972) estimates plant costs 20 percent higher. His figures would lead to a cost of over $32 billion to replace projected imports of gas, and more than $110 billion to fill all our estimated gas needs.

What would be the cost of a full program to synthesize all our projected gas and oil needs? Easily over $100 billion. This figure (which is on the low side) represents only the cost of plant construction. It does not include the costs of extracting and transporting the coal and operating the plant.

What about air pollution for the entire closed system? The conversion of coal into gas will entail the generation of more pollutants. If we draw one-half of our projected total energy needs for 1985 from coal at an optimistic conversion efficiency of one-third (either by generating electricity or synthetic fuels) then we must mine at least 2,400 million tons of coal per year, compared to 600 tons in 1970.

The gas will burn cleanly, but the conversion plant is another story. It may or may not produce air pollution. However, thermal pollution is inevitable. Consider the following excerpts from *Environmental Science and Technology,* which warns us of the possible environmental repercussions of coal gasification.[17]

> From an environmental viewpoint, land, water, and air are all critically involved. One hundred twenty coal gasification plants would more than double present coal consumption. Strip-mining is expected to be utilized to a large extent. The greatly increased amount of mining will require restoration of mined areas.
>
> Water management is also an important factor. For each plant, 100,000 gal/min would be circulated, of which 20,000 would be consumed. Special precautions are necessary to prevent discharging water-soluble contaminants such as phenols.
>
> Thermal pollution in coal gasification is significant since the conversion of coal to pipeline gas is only about 65% thermally efficient (120 plants would require disposal of about 5×10^{15} Btu/year [i.e. our thermal release would more than double.]). Heat disposal to the air is planned by evaporative cooling which consumes most of the water mentioned above. Disposal of coal ash from the coal processed is also necessary. Fortunately, coal gasification is carried out in closed vessels which prevent air pollution.

One hundred twenty plants would still leave us short of our future total gas needs. We should pause to ponder the fact that we would need 500 coal gasification plants (of 90 billion cubic feet per year capacity) to synthesize our total projected 1985 gas needs of 46 trillion cubic feet per year.

DIRECT USE OF COAL

At present, approximately 50 percent of coal, 17 percent of natural gas, and 13 percent of petroleum are employed for the generation of electricity.[18] The question could be raised, "Why not convert all the electricity generating power plants to coal, and thereby save the oil and natural gas that at present go for this purpose?" Could we not reduce the need for synthetic fuels and simultaneously close the energy gap? From the above percentages, by burning coal to generate electricity, we could save approximately 3.7 trillion cubic feet of natural gas and 0.7 billion barrels of oil per year at our present consumption levels. Although these amounts are significant and should not be discounted offhand, they would still fall far short of closing the energy gap. By the mid-1980s, when the oil gap is anticipated to be in excess of 6 billion barrels per year and the gas gap 13 trillion cubic feet, their direct replacement by coal to generate electricity would save us 1.3 billion barrels of oil (approximately 17 percent of the oil gap) and 3.8 trillion cubic feet of gas (about 29 percent of the gas gap). The trouble is that all power plants could not be physically converted to the use of coal, and, for those that could, the conversion would probably come too late and would be too costly. Moreover, as we have seen, coal is a dirty-burning fuel. However, two recent developments show considerable promise along these lines: magnetohydrodynamics (MHD), and a "fluidized bed" process.

The Russians have pioneered in the development of MHD, a technique for the combustion of powdered coal with other additives so that the resulting "torch effect" (combustion) can be used to generate electricity directly.

MHD is only theoretically more efficient than a coal-burning power plant (50 percent compared to about 33 percent). However, a power plant operating on the MHD principle would have few moving parts and would therefore be easily serviced.

The second technique for the clean combustion of coal is the fluidized bed combustion process, in which powdered coal and other chemicals along with air are injected into a molten bed of iron. The iron acts as a heat transfer medium to generate steam to produce electricity. All the impurities in the coal are trapped in the molten iron and form a slag that floats to the top. This technique is still experimental and is designed for the large-scale combustion of coal, as needed in a power plant. Its efficiency and pollution abatement capacity, while promising, are yet to be precisely determined.

SYNTHETIC FUEL FROM ORGANIC MATTER: UNPROMISING

Man could never grow enough organic matter (to convert by fermentation or chemical technology) to produce a combustible liquid as a replacement for gasoline. Besides a negligible energy return, the land that would be required to grow all the organic matter is needed for agriculture, and conversion of organic matter to fuel would yield huge energy losses in the process.

The chart below gives some idea of what we can expect from organic matter power:

SYNTHETIC OIL FROM:	COULD PROVIDE OUR PRESENT NEEDS UP TO:	FOR APPROXIMATELY:
Half of our coal	100 percent	150 years
Organic growth (on arable land not now in use)	10 percent[19]	indefinitely

THE MYTHS OF OIL SHALE AND SEA FLOOR RESOURCES

Let us consider two over-romantic sources of fossil fuels before going on to alternative power sources.

What about the resources of the ocean floor?

There are none.[20]

The deposits of oil and gas that are being discovered "at sea" are taken from the continental shelves, not in the deep ocean basins. They are under a few hundred feet of water in rock strata that were above sea level millions of years ago. Fossil fuels are the partial-decay products of organic matter (plants) that once grew above sea level, or the compacted ooze that is the remains of sea algae that previously grew over a continental shelf area. No fossil fuels are expected to be found on the deep ocean basins. These rocks were never above sea level nor were they ever part of a continental shelf structure. They never had plant growth on them, or algae growth over them, and, hence have no fossil fuel possibilities.

What about oil shale? An unlikely source, as we have seen in the previous chapter. Not only is it difficult to extract and gives a poor return but the hydrocarbon content is rich in nitrogen, which presents refining problems, as it requires more time and expense to remove the nitrogens. With the mention of oil shale, there are three factors to keep in mind: (1) Most shales (99.2 percent)[21] hold about 25 gallons or less of potential oil per ton of shale; (2) although the hydrocarbon content of all shale is estimated at 1,000 trillion "barrels," only one part in 10,000 is recoverable;[22] and (3) oil from shale may add only another 10 percent to man's oil reserves.

The proposal to use nuclear explosions to free the hydrocarbons, as we saw with natural gas, is equally unworkable for the same reasons.

Despite obvious drawbacks, we can see that only coal holds propects for saving the nation from an immediate energy shortage of critical proportions but that the synthetic fuels cannot be produced fast enough on present schedules to close the energy gap of the late 1970s and the 1980s.

What we do now about the attending desecration of the landscape by strip-mining to obtain the enormous quantities of coal needed for conversion into gas and oil, and the pollution from a tenfold increase in coal mining over the next decade, and the multi-billion dollars of capital outlay over the next few years will decide how effectively we can engage in the urgent business of saving our total civilization.

We must accept the fact that the synthetic fuel strategy does not provide "long-term" power benefits and that it brings much "short-term" grief. The energy crisis is deepening daily, and will become incredibly complex by the mid 1980s, when we will be sixty percent dependent upon foreign sources for oil, and vying with other nations for this same foreign oil.

NOTES

1. "Feds Eye Regulations for Strippers," *Environmental Science and Technology* 6 (January, 1972), p. 27.
2. James Branscome, *New York Times,* sect. 6, December 12, 1971, p. 30. © 1971 by The New York Times Company. Reprinted by permission.
3. Ibid.
4. "Feds Eye Regulations," *Environmental Science and Technology,* p. 27.
5. Ibid.
6. Ibid.
7. Ibid.
8. Committee on Resources and Man, National Academy of Sciences, National Research Council, *Resources and Man* (San Francisco: W. H. Freeman and Co., 1969), ch. 8.
9. Ibid.
10. Author's estimations based on population and extrapolated energy rates predicated on general historical data.
11. "Gas From Coal, Fuel of the Future," *Environmental Science and Technology* 5 (December, 1971), pp. 1178-1182.
12. "U.S. Energy Outlook, An Interim Report of the National Petroleum Council," vol. 1, July 1971.

13. "Gas from Coal," *Environmental Science and Technology,* pp. 1178-1182.
14. Ibid.
15. Ibid.
16. Ibid.
17. Volume 5 (December, 1971), p. 1183.
18. Based on data from: *Energy and Power, Scientific American* (September, 1971), pp. 37-49.
19. Based on the productivity of land and the energy equivalency of organic matter after due consideration has been given to losses inherent in converting plant matter into a fuel.
20. Committee on Resources and Man, *Resources and Man,* ch. 7.
21. Ibid., p. 200.
22. Ibid.

The Eternal Sources of Power: Inaccessible or Insufficient

Science-fiction writers, even some scientists, are accustomed to portraying the sun as our ultimate power bank. Consequently, many people have begun to believe that harnessing the sun is just around the chronological corner. This is a misassessment that could be fatal is we let it lull us into a false sense of security. Sunshine is nowhere near being tappable on a sufficiently massive scale for even a fraction of our power needs. Even if it were, we are nowhere near capable of deploying its electrical output. We don't have the transmission lines, the capability of nighttime deployment, and the many satellite industries necessary to deliver it where it counts. Nor will we have the *capacity* to make electricity serve us in the foreseeable future, as we will see in a later chapter on copper, the conductor of electricity.

Indeed, sunshine may be our ultimate power source, but America could collapse while dreaming about it in philosophical detachment from practical matters.

Additional eternal power sources are those natural phenomena that will last as long as the sun and the earth – flowing water, wind, tides, earth heat (the last being, technically speaking, a non-renewable source). All these sources, if harnessed by man to any significant degree, could only be coaxed into giving us some of their power in the form of electricity. Even though each of these sources is immortal, the power that each could deliver is either quite insufficient for our present needs or it is inaccessible to our technological grasp.

SOLAR POWER: A GIANT BEYOND REACH

The energy of the sun is delivered to the earth in the form of radiation. Sunlight is a form of radiant energy. Any form of light, as the light from a light bulb, is also radiant energy. Other forms of radiation are visible light, radio waves, infrared light, x-rays, gamma rays, and so on.

Photons make up the sunlight (and all radiant energy). They possess no weight; they are "pure energy." There are about one trillion times one hundred thousand photons per square inch per second in sunshine before it enters the earth's atmosphere.

The total energy radiated to the earth by the sun is 180,000 trillion watts.[1]

At the rate of 0.9 watt per square inch, the earth is the beneficiary of 180,000 trillion watts of power.

In other words, the solar radiation reaching the earth is 30,000 times the present total industrial power employed by man (6 trillion watts). Surely if our enormous creative talents and technological skills could harness such vast potential power, our industrial life-support systems would last as long as the sun.

Unfortunately, this prodigious total power input to the earth is dispersed over an immense area, so that the power input density is very low. Whereas the power density at the earth's outer atmosphere is 0.9 watt per square inch, the power density at the surface of the earth is much lower.

Of the total incoming solar radiation to the earth, about 46 percent is absorbed by the atmosphere and earth, about 30 percent is reflected, about 23 percent causes evaporation of water from the seas, and about 0.2 percent drives the winds – only about 0.05 percent is used by green plants in photosynthesis.

A power station would need to have an enormous radiation collection area to gather sufficient total power from the sun's low-density radiation. The following calculation illustrates this.

What size solar panel (or concave mirror to focus sunlight) would be needed at New York's latitude to power an average-size power plant (1 billion watts)? The average energy from the sun absorbed at New York's latitude at ground level is about 0.15 watt per square inch. This is only 17 percent of the sunlight's maximum power in outer space. Assuming a 10 percent conversion efficiency, we would need a panel about 7 feet by 7 feet to light a 100-watt bulb.[2] To provide 1 billion watts of power, we would require a panel, if shaped as a square, *about 4 miles on each side!* To provide electrical power for New York City would require a solar panel 15 miles on each side. Although the collecting devices obviously could not be located in the metropolitan area, it would require about ten times the land area of Manhattan. All these calculations based on average solar power reception are on the extremely conservative side. We have based them only upon present needs. When we run out of gas and oil and must depend upon electrified ground transportation, the problem will be magnified many times.

In considering solar power as a source of electricity we must also consider the storage of electricity generated during daytime for nighttime use.

Placing the solar panels in space (23,000 miles high in synchronis orbit), as some suggest, may solve the cloud cover problem but would not obviate the panel's 10 percent efficiency. Some experts theorize that panels of a 50 percent efficiency are possible. Perhaps one day they will be developed. However, the total mass of all the electrical equipment to be positioned in space would run into the millions of tons. This would require tens of thousands of rockets, each equivalent in power to one Atlas rocket. Another cause for concern is the effect of micrometeorites on the panels over the decades. This could be a major and fundamental drawback to this plan's reliability.

Even assuming that all these problems can be solved, what of the other requirements of deploying solar power?

If our electrical needs by 1980 rise to about 500 billion watts, from the present 375 billion, then we would need about 8,000 square miles of sunshine-collecting surface.[3,4]

Solar cells (made of silicon) are absolutely out of the question; there simply is not enough silicon semiconductor industry to do even a small fraction of the job. If we used mirrors to collect the sunlight, we would need our entire 1970 production of aluminum (10 billion pounds) to be rolled out into reflector sheets for the parabolic mirrors to about 1/200 of an inch – or paper thin.

Estimates for the cost of electricity from solar cells are for about 500 times the present cost of electricity.[5] This could be brought down by mass production economics but not to the point where electricity would be as desirable, or practical, as it is today.

Nor can 500 billion watts of electrical power be used directly and immediately. How can we get solar power at night? Two suggestions are frequently mentioned. The first is to electrolyze water into hydrogen and oxygen gases by passing an electric current through it. The resulting hydrogen and oxygen gases can then be stored for later use. This is called the secondary fuel route. They can be burned in a turbine to generate electricity when needed. Such a scheme would entail massive amounts of secondary equipment and a great loss in energy during the various conversion processes. This would entail the need for even greater panel (or mirror) areas to collect enough sunlight in the first place.

Another suggestion to deploy solar power for nighttime use is to use it during the day to pump water from a river uphill to a reservoir so that at night the released water can generate electricity by flowing through a turbine – just as hydroelectric power is generated.

The secondary fuel route, as a supplier of all our electrical needs by solar power, will entail the creation of an industry with a physical plant capacity about half as large as our entire oil industry. Deploying solar power through the hydroelectric-like proposal also requires an enormous reservoir and pumping capability.

Other work under way in solar power harnessing includes: (1) the use of flat plates of photovoltaic cells (used in spacecraft for heating water and generating electricity), which has an efficiency of about 10 percent, (2) the utilization of a hothouse effect on ducts carrying molten sodium and potassium that produces temperatures of about 1,000° Fahrenheit – becoming the heat-exchange medium for a steam-electric power plant at a conversion efficiency of about 30 percent, and (3) a boiler system heated by the focusing of sunlight by mirrors, in which steam is generated to produce electric power at an efficiency of about 20 percent.[6]

However, even these most promising systems require very large areas for the sunshine-collecting devices. It is here that these systems become impractical – at least for our present state-of-the-art in engineering.

Costs of orbiting solar panels are anywhere from $50 to $1,000 per pound of weight placed in orbit.[7] To position in space, each 10-billion-watt station would cost between 250 and 5,000 million dollars. This is just the rocket-launching cost. For one trillion watts of electricity from orbiting solar stations the cost would be 25 to 500 billion dollars, merely for launching the rockets.

WIND POWER

The wind will blow as long as the sun shines. However, it is even more dispersed than solar power. Since sunshine drives the wind, by uneven heating of the atmosphere, wind power potential must be less than solar power potential; that is, less than 0.9 watt per square inch. Actually it is approximately 20 percent of all solar power, but its dispersement covers the entire area of the earth, and is expended throughout the entire atmospheric volume.[8] Hence, to tap wind power would require windmill-sail areas much larger than solar panel areas for equivalent amounts of electrical power. Also, the steady winds blow high in the atmosphere so that towering windmills would be needed – about one to ten miles high, or between five to fifty stacked Empire State buildings.

Just as with solar power, wind power would be tapped as

electricity. The same problems of deployment arise with respect to this source.

FLOWING WATER – OR HYDROELECTRIC POWER

Hydroelectric power is generated by damming rivers to create a lake at a sufficient depth for the lower lake waters to be under sufficient pressure to spin an electricity-generating turbine. For the sole purpose of creating electricity, with no consideration given to agriculture, it would require damming all major rivers just to produce one-half (3 trillion watts) of the world's present total power expenditure. Only 5 percent of the total potential of 3 trillion watts has been developed as of 1965.[9]

The following table shows hydroelectric power potential by region:[10]

REGION	PERCENT OF TOTAL POTENTIAL (3 trillion watts)
Africa	28
South America	20
U.S.S.R. and China	17
Southeast Asia	16
North America	11
Western Europe	6
Australasia	1
Far East	1
Middle East	negligible

Current total United States electrical power output is 375 billion watts. Our entire hydroelectric power potential is 160 billion watts. Hence, we cannot obtain even our present electrical needs from this eternal source of power.

If we compare the countries with the greatest hydroelectric power potential with their populations we get a maximum living standard derived from hydroelectric power alone. Of course, such hydroelectric power is a long way from manifesting itself as houses, cities, transportation, schools, and so on.

Note in the chart below that Norway alone, because of its small population, has a higher living standard horizon than the United States and we can see that potential wattage is made meaningful only by reference to population.

With the United States at the bottom of the list, if we were suddenly bereft of other sources of power, our living standards would be reduced to those prevailing in such impoverished nations as India and Pakistan. Alternatively, if we were to attempt to maintain our overall living standard (10,000 watts per capita) by hydroelectric power alone, we would have to reduce our population to 16 million.

LIVING STANDARD HORIZONS
FROM HYDROELECTRIC POTENTIAL[11]

COUNTRY	WATTS PER CAPITA FROM HYDROELECTRIC POTENTIAL
Norway	13,000
Africa	4,000
U.S.S.R.	2,200
Canada	1,300
United States	800

Fortunately, the generation of electricity by water power produces no air or water pollutants. However, there are other considerations.

The damming of rivers creates lakes behind the dam. On many sites these lakes serve as reservoirs for irrigation of the land. On other sites the lakes flood the land. The theoretical hydroelectric power potential of the world is calculated without regard to this issue. Hence, the hydroelectric power *worth* developing is less than the 3 trillion watts estimated as the world's potential.

For example, the Aswan Dam has had a surprising and totally unanticipated effect upon the local environment. Actually, the effects were not totally unanticipated. Dr. Abdel Azez Ahmed, a noted Egyptian hydrologist, had warned of some of the dangers which have since come to pass. However, after voicing his objections, he was fired by the government in Egypt.

For eons the Nile River has flowed from the interior of Africa to the Mediterranean Sea, carrying with it suspended matter rich in natural nutrients, the rich soil gathered in its watershed – the Sudan. Seasonal floods have deposited these nutrients along the banks of the Nile, enriching the farm lands and providing food for the populace.

With the construction of the Aswan Dam the Nile waters were gathered into a lake – Lake Nasser. The lake water, being less turbulent than the freely flowing river, lost its suspended matter as sediment to the lake bottom. The lake water, released from the dam, does not flood the Nile basin as in centuries past; it's sterile, deprived of its nutrients.

Hence, the Egyptian farmers must now use fertilizers to achieve a yield twenty percent less than in the past. Considering that Egyptian farmers earn an average of seventy-five dollars per year, this loss could be catastrophic to their already impoverished living standard. Moreover, the Nile itself does not support the algae growth it used to, which explains why the fish catch has dropped in the Nile Delta.

There is also a serious health hazard that has already taken on catastrophic proportions. The snail carriers of a severely debilitating disease, schistomiasis, have always lived along the banks of the Nile. However, in years past the annual floods have washed many of these snails out to sea. They require placid waters to establish a foothold and multiply themselves. Lake Nasser now provides these quiet waters. In four months, these snails can multiply themselves 50,000-fold. These carry the larvae of a blood fluke which, when they enter a human host, steadily weaken him by attacking many vital organs of the body (liver, stomach, heart, and lungs). An individual so afflicted can rarely work more than a few hours a day, and many die. Although comprehensive data are not available concerning the outbreak of schistomiasis since the construction of the Aswan Dam, it is known to have jumped from zero to eighty percent in some areas.

The dam itself is not providing the hydroelectric power that was forecast for it because Lake Nasser is only half full. It had been assumed that the Nile, flowing into Lake Nasser, would seal up the lake's sandstone bed, eliminating the loss of water through ground leakage. This has not occurred. Instead, the waters in Lake Nasser seep into the sandstone. Coupled with natural evaporation, Lake Nasser, which was expected to be full by now, is filling up at an unforeseen slow rate. With less water comes less hydroelectric power. Presumably, the dam's ultimate advantages in Egypt will outweigh its disadvantages, but a principle emerges: the subleties of a dam's effects are sometimes far-reaching.

TIDAL POWER

The motion of the tides dissipates some one to three trillion watts of power, one-half our present world power output.[12] A great share of this "tidal friction" energy is dissipated over the continental shelves. Some calculations indicate that the tidal friction of all the seas on earth is slowing the earth in its rotation by about 0.00001 seconds per year. This would mean that when multicellular life began in the seas some 600 million years ago, the earth had a day of approximately 23 hours.

Even though all the tidal power, if harnessed by man, would be sufficient for half of his present power needs, an apparently insurmountable problem is that this power is too dispersed along the beaches of the world to be tapped.[13] On most of the beaches, the average tidal height (three feet) is insufficient to create high-enough pressure to yield meaningful amounts of electricity. For electric power generation, the tides should have a depth of approximately thirty feet and the tidal basin should hold sufficient sea water to generate current for hours.[14] Very few beaches meet these needs, where the contour of continental shelves gathers enough water at high tide at sufficient depth to warrant the construction of a power plant.

All developable sites for tidal power therefore would yield only 1 percent, at best, of all of man's present power needs.[15]

Whereas the yield is negligible, however, possible adverse effects on the environment through the development of tidal power would be minimal. There would be no air pollution and no foreseeable derangement of local sea life.

GEOTHERMAL ENERGY: A SUBLIME BARRIER TO BREACH

Can the heat within the earth itself be tapped to generate steam, which, in turn, could be used to generate electricity? Theoretically, as the earth's heat is converted into electric power, the total heat in the earth would drop. However, the heat reservoir of the earth is so large that, were it tappable, the rate of "cooling" would be incredibly slow. This quantity of energy can be considered inexhaustible in a practical sense. It is, very approximately, enough to supply our total industrial energy needs for several million years. Considering the earth's heat as an energy bank, it is about 100,000 times all that of the fossil fuels. But it is less than one percent of that for fusion power potential, which we will cover in a later chapter.

However, the heat of the earth is either conducted to the surface or convected to the surface.

The heat that is conducted to the surface appears as normal ground temperatures a few feet below the surface. This low temperature and low heat flow preclude conversion into electrical power on thermodynamic grounds alone.

The heat that is convected by lava flows to the surface appears in only a few places on earth. There are very few sites on earth where hot rock upthrusts bring a sufficient lava flow and hence a sufficient heat flow to the surface to warrant the construction of a power plant. Yellowstone National Park and hot-spring areas of New Zealand and Iceland are better-known sites.

Furthermore, drilling into the earth's mantle, some thirty miles down, is very impractical. The deepest well ever drilled is only seven miles deep. A gradual rise in the plasticity of the crust at increasing depths would seal off any well, even if one could be drilled so deep. The major barrier to the earth's heat is the very gradual rise in temperature of the earth's crust at increasing depths. This prevents the proper heat flow characteristics required for power generation.

Using the heat of the earth to drive steam turbines appears to be achievable only on a very small scale in a very few selected places where there is a sharp temperature gradient instead of the usual one-degree rise per 100 feet of depth. Thus, although the total geothermal potential is great, only a small portion of this is accessible. The portion that is technologically tappable is loosely referred to as our geothermal potential. This fraction, if fully developed, would generate only about 1 percent of present world total energy needs.[16]

Nor must the possible adverse effects of geothermal power development on the environment be automatically assessed as zero. Tapping hot rock upthrusts as heat sources by pumping down water to obtain steam for steam-driven turbines for electric power generation may, over many decades, cause density changes in the rock strata and hence alter its stability against earthquake-like movements.

Also, some steam emitted by geological sites is rich in hydrogen sulfide gas – a very serious air pollutant. Unless the hydrogen sulfide can be removed from the steam this scheme will cause air pollution.

Geothermal energy is beneath our feet, but out of our technological grasp – at least to the extent of more than one percent of our total energy needs. Not everyone agrees.

There is a novel plan to tap geothermal energy.[17] This scheme would not rely upon natural geysers, as do the present geothermal power plants in California, New Zealand, Iceland, Japan, and Italy. Nor does it call for drilling into the earth's mantle. Rather, scientists at the Los Alamos laboratory conceive that it should be possible to drill down about five miles into hot rock upthrusts with standard oil-drilling equipment, then pump down water at 7,000 pounds-per-square-inch pressure to cause fracturing of the rock – a technique called hydraulic faulting. A large underground pancakelike fracture zone would result. Next, another well would be drilled about thirty feet from the first. Water would be pumped down the second well, which would circulate and become heated in the fracture zone and obtained as hot water from the first. This hot water, under high pressure, could then be used to power a steam-electric power plant.

Computer analysis of the earth's heat flow characteristic indicates that each well could supply energy for about thirty years for a 100-megawatt power plant. Extracting energy will cool the heat source. The number of suitable geothermal sites is not yet known, but according to the Los Alamos scientists there should be enough for all our electrical needs.

There are several dimensions worth noting in this proposal. A lifetime of 30 years for a well means a 3 percent per year turnover in power-plant relocating. We have seen that by 1985 we shall probably need 750 billion watts of electrical power. If only 10 percent of this came from 100-megawatt geothermal power plants we would need 750 such plants, and every 30 years they would have to be relocated.

If these power plants actually work as the computer-run shows, it will nevertheless take at least a decade (not the full 30 years) to prove this; hence, it will take many decades before this scheme can supply us with large percentages of our electrical needs. This is not a drawback. The plan should be pursued vigorously. But we shouldn't expect this scheme to close the electricity gap of the 1970s and 1980s. Perhaps it can come to our rescue after the year 2000, just as in the case of solar power development – that is, if it's really workable.

ETERNAL POWER SOURCES
AND WORLD ENERGY NEEDS

SOURCE	APPROXIMATE POWER POTENTIAL	BUT TECHNO-LOGICALLY
Hydroelectric	1/20 present needs	insufficient
Tidal	1/100 present needs	insufficient
Wind	100 times present needs	inaccessible
Solar	30,000 times present needs	inaccessible
Geothermal		
tappable heat flow:	1/100 present needs	insufficient
total energy bank:	100,000 times fossil fuels	inaccessible

Obviously, eternal power sources are beyond our grasp; fossil fuels of oil and natural gas are running out; coal cannot be converted into oil and gas fast enough; dependency on foreign fuel is impolitic; and we are unlikely to elect fuel rationing, or power reductions, each of which would lower our living standard drastically.

Apparently, we have no choice; we must turn to nuclear power. But, even here, is nuclear power the answer?

NOTES

1. Committee on Resources and Man, National Academy of Sciences, National Research Council, *Resources and Man* (San Francisco: W. H. Freeman and Co., 1969), ch. 8.
2. Ibid.
3. U.S. Atomic Energy Commission, *The Nuclear Industry* (Washington, D.C.: Government Printing Office, 1971), p. 9.
4. Committee on Resources and Man, *Resources and Man,* ch. 8.
5. Glenn T. Seaborg and William R. Corliss, *Man and Atom* (New York: E. P. Dutton and Co., 1971), p. 55.
6. *Energy and Power, Scientific American* (September, 1971), pp. 61-70.
7. Seaborg and Corliss, *Man and Atom,* p. 55.
8. *Energy and Power, Scientific American* (September, 1971), p. 61.
9. Committee on Resources and Man, *Resources and Man,* ch. 8, p. 209.
10. Ibid.
11. Authors' calculations based on data in this chapter.
12. Committee on Resources and Man, *Resources and Man,* ch. 8.
13. Ibid., p. 211.
14. National Academy of Sciences.
15. Committee on Resources and Man, *Resources and Man,* ch. 8.
16. Ibid., p. 218.
17. "New Plan Is Outlined for Tapping Geothermal Energy" *New York Times,* June 21, 1972.

Our Atomic Energy Program: Too Little Too Late

Our atomic energy program is a classic example of too little too late. Not until about the year 2020 will it be capable of supplying us with all our electrical needs.

The awareness will eventually dawn that even nuclear fission (uranium) energy is a temporary source of power – and cannot even figure into our immediate problem of the next thirty years.

These are the shortcomings of fission power – not its environmental hazards, as too many people now erroneously believe.

To contemplate the energy crisis in the United States is to be drawn to the bitter ironies of our past indifference and neglect.

Coal is our dirtiest fuel. To obtain it, we are forced to rape the landscape. Yet coal appears to be our only hope of averting a closing economic, political, and social catastrophe.

Nuclear energy was recognized in the forties as the power of the future. However, it has been so thwarted by those concerned about its environmental consequences that it has moved at a snail's pace. It cannot possibly deliver significant amounts of electrical energy before the turn of the century. In short, nuclear energy cannot rescue us from our present plight.

One of the great concerns of the sixties was to liberate the inhabitants of the ghettos. We aspired to create "The New Frontier" and "The Great Society." For the concern of the seventies and eighties, all roads point to the question: "How do we prevent all America from becoming one giant ghetto?"

After World War II, the United States was magnanimous in sharing its vast resources with friend and foe alike. By the 1980s, we may depend upon Russia and the Middle East to be magnanimous with us. For, if we fail to accelerate coal gasification now and continue to dawdle in the nuclear energy field, a reverse Marshall Plan may be our fate.

The greatest irony of all is that the sincere advocates of opposing power schemes are so polarizing the nation that no scheme is likely to succeed. Nobody has declared the real issue: *The rate of development of any of our power options is frighteningly behind any schedule that will prevent a power blackout in American in the late 1970s.*

Whereas this chapter is, in a sense, about a lost opportunity, it is also about a present and future challenge – nuclear fission, the splitting of the nucleus of an atom into nuclei of lighter atoms, which releases energy.

What is energy and how does it differ from matter? Energy, we have seen, is the power to do work, but it is also the cause of all changes that occur in matter, and it is intangible. Matter is anything that can be weighed, or seen, and is tangible.

All matter is composed of atoms. Atoms are extremely small – about one billionth of an inch. This page is about 1,000,000 atoms in thickness.

The 92 different atom types in nature are called elements. Each atom has a center, or a nucleus, occupying a space of about what a pinhead would on a ball 10 feet in diameter. But the nucleus comprises about 99.99 percent of the atom's mass.

Within the nucleus are protons (positively charged bodies) and neutrons (no charge). About the nucleus are electrons (negatively charged bodies) – one for each proton in the nucleus.

The nucleus of any one element is variant; there are several varieties of nuclei, each with its own specific number of neutrons.

These nuclear variations of an element are called the isotopes of the element. Some isotopes are radioactive, others are not.

THE URANIUM ATOM: THE SOURCE OF ATOMIC POWER

Uranium exists in two principal isotopes, U(235) and U(238). Their relative abundancies in nature are:

U(235) – 0.7 percent of all uranium

U(238) – 99.3 percent of all uranium

The energy stored within the nucleus of an atom is the potential energy of the protons and neutrons. Uranium has a tendency to cleave, or fission, and thereby produce atoms of lower atomic number and lower atomic weight but of more stable nuclei. This process releases energy. This tendency to fission is measured by uranium's natural decay rate, called the half-life, which is defined as the time that must elapse for one-half of the nuclei to decay. Uranium's half-life is about 4.5 billion years. That is, a pound of uranium will lose half of its radioactivity in 4.5 billion years, half of the remaining radioactivity in another 4.5 billion years, and so on.

All the uranium on earth was created with the creation of the universe and has been decaying ever since. Uranium therefore is a nonrenewable resource. Given enough time, billions of years, all the earth's uranium will decay naturally and all its stored energy will have been released very slowly over the full span of time. But man can accelerate this energy-release process. After mining uranium ore and extracting the uranium as a metal, scientists can induce the release of any specific quantity of this energy in a millionth of a second of a bomb explosion or the same quantity over centuries in a power plant.

The induction of accelerated decay is accomplished as a consequence of a "chain reaction." The chain reaction for uranium is:

Uranium + 1 neutron → other atoms (40 isotopes) + 2.5 neutrons (average) + energy

One uranium nucleus absorbs a neutron and then fissions to yield lighter atoms plus 2.5 neutrons (average figure) plus energy. Each uranium nucleus that fissions can induce the fission of at least two other uranium nuclei. This is called the "chain reaction."

The nucleus of uranium fissions spontaneously (naturally) – it does not have to be made to fission. Rather, the rate of fission for a given uranium sample can be speeded up or slowed down.

To construct a bomb, the uranium would be fashioned into special shapes so that when the metal shapes are suddenly brought together to form a sphere, about the size of a grapefruit, the chain reaction occurs – in a millionth of a second throughout the total mass.

The secret of the atomic bomb was how to bring together uranium metal shapes in the briefest time to achieve the "critical mass." This operation is called the trigger mechanism. Its details are classified information.

In the construction of a power plant, the uranium is in the form of an oxide and contained in pellets about the size of a thimble. These pellets, by the millions, are encased in steel tubes many feet long and an inch or so wide. The total mass of these pellets is approximately 100 tons (for a 1,000-megawatt power plant). Thousands of the tubes, or rods, are inserted into a reactor's core. This arrangement, with the rods absorbing neutrons from the uranium fuel in the pellets, controls the rate of induced nuclear decay. When the nucleus of uranium fissions (splits), the energy released appears as heat. The heat is used to generate steam. The steam is used to drive a turbine. The turbine generates electricity.

The energy transformations are: nuclear potential energy to heat; heat to mechanical energy; mechanical energy to electricity. The total efficiency of an atomic power plant is 30 percent. Thirty percent of the nuclear potential energy appears as electricity that the power plant puts out; the other 70 percent is waste heat. An important technical point is that 1.2 kilograms of uranium can produce one megawatt of electricity for one year; or 1,200 tons of uranium can produce 1 trillion watts of electricity for one year.[1]

A NUCLEAR POWER PLANT (SCHEMATIC)

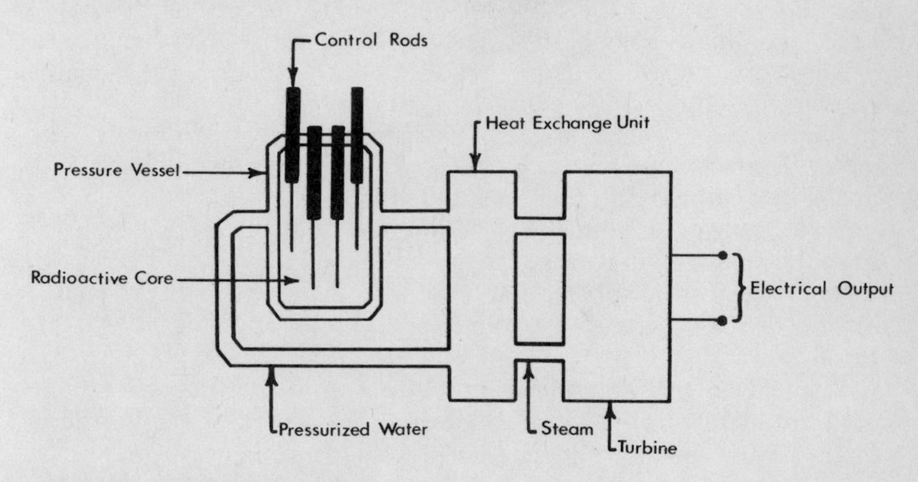

NUCLEAR POWER PLANTS
IN THE UNITED STATES (1972)[2]

STAGE OF DEVELOP-MENT	NUMBER	POWER OUTPUT IN MEGAWATTS
Operable	23	10,040 (10.0 billion watts)
Being built	54	45,780 (45.7 billion watts)
Planned	52*	51,570 (51.5 billion watts)
Total	129	107,390 (107.3 billion watts)

*Only 13 of these have received construction permits.

When all the plants listed above are fully operative, they will provide only 29 percent of our *present* electrical needs, or 107 billion watts to meet our expenditure of 375 billion watts. If all 129 nuclear plants are operable by 1985, they will supply only 15 percent of our projected 750 billion watts of electrical power.

The reactors that are now operable are called "burners"; they use as a fuel uranium(235). The fuel pellets contain uranium(235) to the extent of 3 percent by weight. Natural uranium contains only 0.7 percent of the 235 isotope by weight. The process that creates fuel pellets of 3 percent uranium(235) is called "enrichment." The only countries presently able to enrich commercial quantities of uranium are the U.S. and U.S.S.R. The technical details of the enrichment process are classified information. Hence, all other countries planning to develop atomic power must purchase their uranium fuel from Russia or America. This situation will no doubt change either as new reactor designs that don't need enriched uranium are developed or as European consortiums build and perfect the enrichment technology.

THE "BREEDER" REACTOR: POWER PLANT OF THE FUTURE?

Of the two isotopes of uranium, U(235) and U(238), only U(235) is readily fissionable by slow neutrons. Upon fission, these neutrons are moving very fast and must be slowed by a moderator (graphite or water) to increase their absorbability by other uranium(235) nuclei. Fast neutrons are preferentially absorbed by uranium(238) nuclei. U(235) is used in bombs and in power plants. However, with an abundance of only 0.7 percent of all uranium, to obtain the other 99.3 percent of potential energy in U(238), the U(238) must be converted into a more readily fissionable isotope – a process called "breeding." The U(238) is converted into plutonium(239), referred to as Pu(239). By an arrangement that permits it to capture a fast neutron, Pu(239) can serve as a fuel just as U(235) can. The breeder reactor would consume U(235) as a start and generate electricity, as with a nonbreeder, or "burner." However, it would also generate Pu(239) from U(238), which would be included into its design. Upon exhaustion of the U(235) the reactor would have produced more Pu(239) than the U(235) it consumed. This Pu(239) then takes over as the fuel to generate electricity. Upon its exhaustion, more Pu(239) will have been "bred" from U(238), added for this purpose, than the Pu(239) that was consumed as a fuel. As long as it is supplied with U(238), the breeder reactor can generate electricity and more fuel than it started with.

THE BREEDER REACTOR [SCHEMATIC DIAGRAM FOR U(238)]

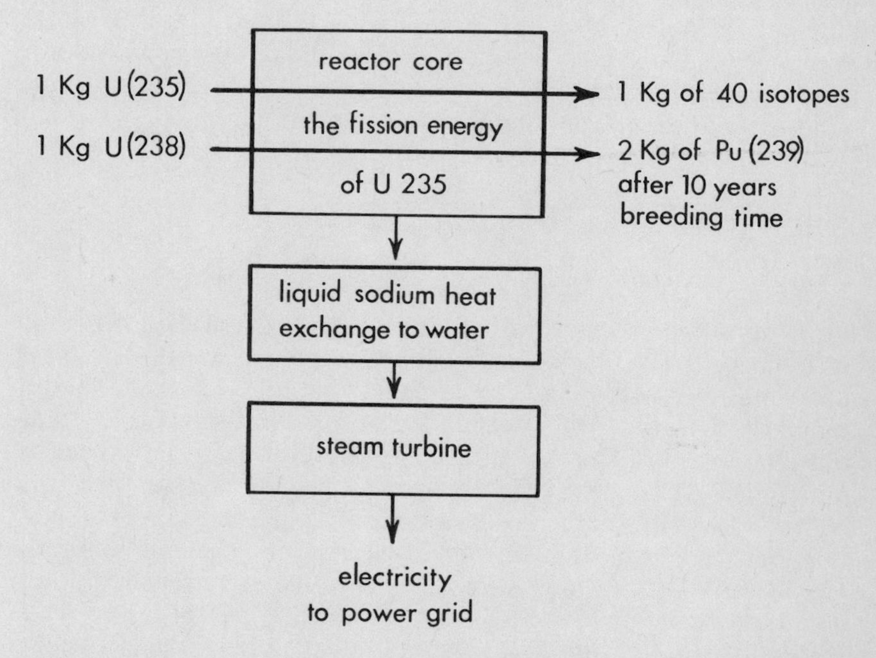

One ton of U(235) fissions to produce electricity, one ton of isotopes, and two tons of Pu(239) from two tons of U(238) in a ten-year period.

The electric power output equals about one-third of the fission power yield. The other two-thirds is accounted for by a heat output (thermal "pollution").

Another advantage of the breeder reactor is that it could convert thorium(232) – Th(232) – into U(233). Thorium(232) is not a fuel, but U(233) is as fissionable as U(235) or Pu(239). This reactor is in a far less developed stage than is the U(238) breeder reactor.

The energy bank of United States uranium is not as large as many people think. The energy equivalent of U(235) is only equal to that of our oil reserves!

If the breeder reactor is perfected and on-stream in the mid-1980s, as the present development program calls for, then the energy bank of U(238) would be tappable. It would at best be double our coal reserves. This more optimistic picture would be due to the fact that the breeder reactor would permit use of poorer grades of uranium ore.

The National Academy of Sciences is strongly in favor of the breeder reactor over the "burner" type of reactors because the burner uses up the uranium(235) isotope. Only the fission of the 235 isotope can provide the neutrons to breed plutonium(239) from uranium(238): it is a precious match without which we cannot light the atomic future of the breeder reactor.

The first commercial breeder reactor will go on-stream in Scotland within a year as part of the United Kingdom's atomic energy program. Our first commercial breeder reactor is scheduled for 1980, according to the AEC.

Estimating uranium reserves that ultimately will be recovered presents the same problems as with all resources. Estimates change according to the statistical approach used and the cost of recovery for low-grade ores. Rifford L. Faulkner, director, Division of Raw Materials, U.S. Atomic Energy Commission, estimates our reserves at 660,000 tons; the free world's at 1.5 million tons.[3] Others estimate United States recoverable uranium at from 200,000 to 3 million tons. But the center of gravity of all estimates is one-half to 1 million tons. We have taken Faulkner's estimate of 660,000 tons as a working figure. These are the figures accepted by the National Academy of

Sciences, whose figures we have relied on throughout.

How long could the uranium reserves of the United States last? Suppose that all power were to be drawn solely from nuclear fission. We'll work with the proved plus estimated reserves of 660,000 tons, a United States power output of 2 trillion watts, and a power expenditure of uranium at 1.2 kilograms (1 kilogram = 2.2 pounds) of uranium per year per megawatt of power. Therefore, to supply our present 2-trillion-watt needs for a year requires 2,400 tons of uranium. At such a comsumption rate, uranium would last for 275 years.

Power output of uranium	=	1.2 kg/yr/megawatt (1,200 tons per year per trillion watts)
Power output of United States	=	2 trillion watts
Uranium needed per year	=	2,400 tons
Total uranium reserves	=	660,000 tons
Lifespan of reserves	=	275 years

Suppose that the present rate of power growth continues until the year 2000. At our present 7 percent per year growth rate we would need 16 trillion watts by 2000, or 8 times our present needs. Apart from the impossibility of our being able to achieve, and sustain, such expenditures of power, one can see from the figures above that we would deplete our entire uranium resources in about 30 to 40 years. Clearly, even the nuclear power of fission cannot sustain mankind for today's expanding numbers and living-standard pretensions beyond one-ten-thousandth the time man has already lived on earth.

THE BREEDING GAP: A SERIOUS PROBLEM

Have we waited too long before beginning the breeder reactor program?

Let us consider two basic facts. All the readily available uranium ore in our nation that could be processed by 1980 amounts to about 210,000 tons.[4] If we could process this ore to create enriched fuel elements for the breeder reactor, we could obtain about 1,500 tons of U(235) fuel at most based on a 0.7 percent content of uranium. This is what we would start with to breed plutonium from U(238). (It would be in the form of 2 to 3 percent "enriched" uranium fuel pellets.)

Secondly, it takes about ten years to double the fuel inventory in a liquid metal fast breeder reactor.[5]

The disheartening conclusion is obvious. Even if we could possibly start with 1,500 tons of U(235) fuel in 1980, we could not expect more than 3,000 tons of Pu(239) by 1990, more than 6,000 tons of Pu(239) by 2000, more than 12,000 tons of Pu(239) by 2010, and so on.

Of course during the ten-year cycle of breeding, we would obtain electricity from the breeder reactor, but only a small percentage of our anticipated needs. In fact, if our electrical need rises to 1 trillion watts from the present 375 billion watts, as the AEC predicts in their booklet "Why Fusion?", and then holds constant, we could catch up to this amount from the breeder reactor only by the year 2015 – if we accept the impossibly optimistic 1,500-ton schedule above.

There is a breeding gap! The breeder reactor program should have begun during the 1940s. The following chart will illustrate the magnitude of the gap. However, the figures are highly optimistic because they assume that all readily available uranium ore can be processed by 1980 to yield 210,000 tons of uranium, and then enriched to give 50,000 tons of 3 percent U(235). Also, we would need about 100 breeder reactors to use this fuel inventory, but we will have only one breeder reactor by 1980 according to present plans.

From the chart we can see that during the years 1990 to 2000 we can expect no more than 25 percent (b) of our electrical power from the breeder reactor program. This is only a small percentage of total power needs. This is truly an energy gap of serious propor-

THE BREEDING GAP (ELECTRICITY)

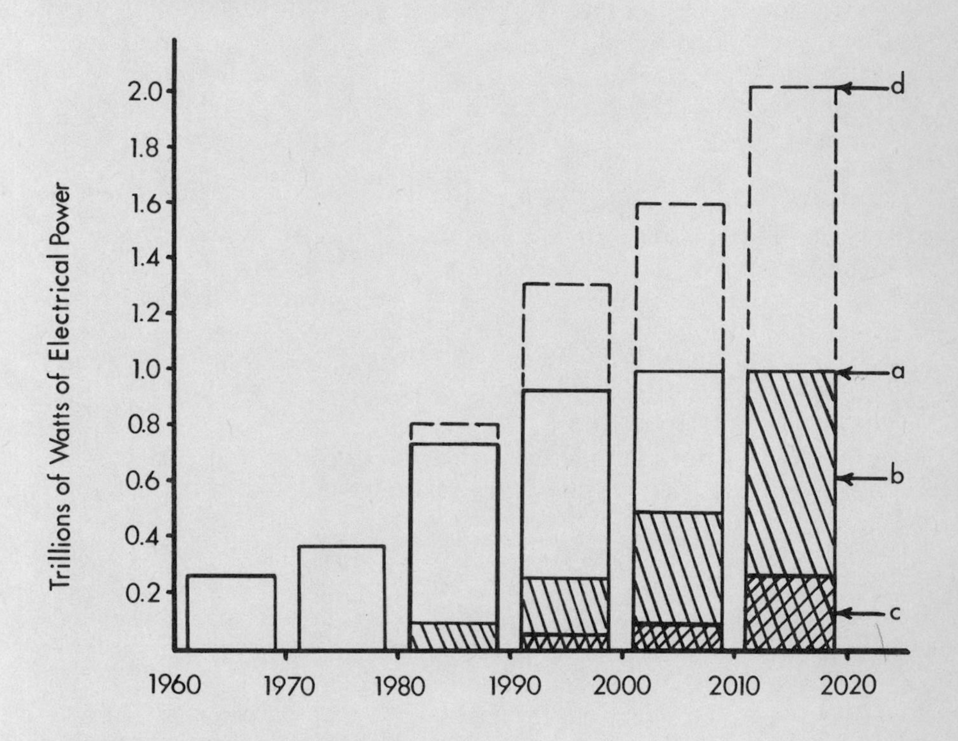

a. Expected electrical needs.
b. Fraction of electrical needs provided by the breeder.
c. A more realistic estimate of our electrical potential from the breeder reactor program.
d. The great range in predicted electrical power needs.

tions, even with our optimistic 1,500-ton figure.

The power gap that is opening is of a great magnitude – about one-third of our total needs, as indicated in the following chart. Should this gap actually materialize, our industry will not simply regress; it will collapse.

U.S. ENERGY FROM ALL SOURCES

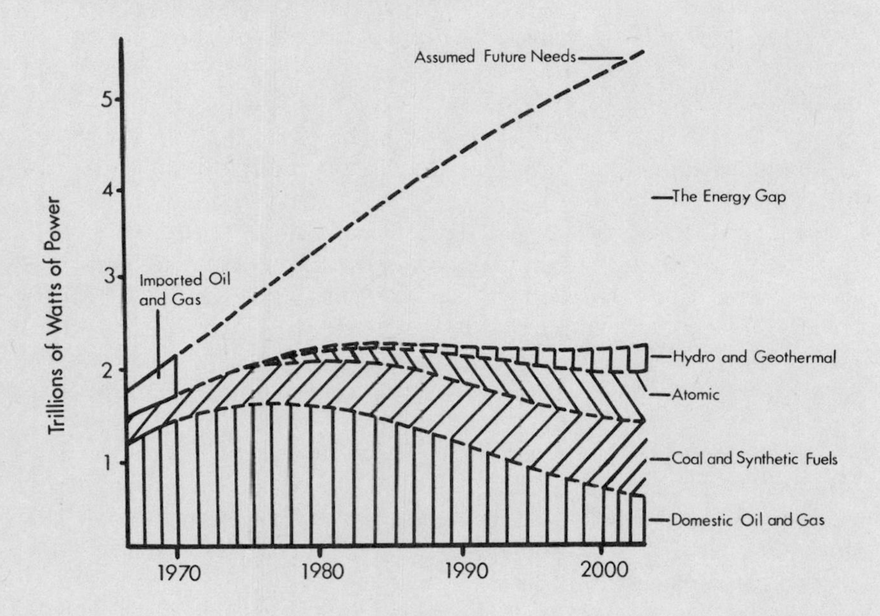

EFFECTS OF NUCLEAR RADIATION ON LIFE

Whether uranium is used in bombs or in power plants, its fission generates a spectrum of some 40 isotopes. Some of these isotopes are radioactive and hence hazardous to man and the environment. What are these radioisotopes and how are they dangerous?

Whereas isotopes have very similar chemical behavior, some are not stable: their nuclei will emit radiation. Radiation can appear in several forms, for example as alpha radiation (a helium nucleus), or beta radiation (an electron), or gamma radiation (a photon). These entities (and others) damage the living cells through which they pass.

Radiation damage to living cells is caused primarily by the fragmentation of cellular water and the subsequent attack of the water fragments (free radicals) on the various parts of a cell. The parts most adversely affected are the genetic code (DNA) and the enzymes. All cells of plants or animals possess enzymes. These are chemical tools that enable the chemistry of life to go forward at ambient temperature (normal cell temperature). Should the enzyme systems be exposed to radiation they lose their effectiveness as chemical tools. The result is the shutdown of the chemistry of life.

Seeds, because they have less cellular water, tend to be more resistant to radiation damage than cells. The species of plants or animals that appeared earliest in the evolutionary sequence of life are more resistant to radiation than those species that evolved later, because of the presence of greater genetic redundancy (greater genetic repetition given by a greater number of chromosomes).

Nature herself is imbued with a background of radiation. This intrinsic radioactivity is caused by the presence of radioisotopes that were created when the universe was created [uranium(238), potassium(40), for examples], or by solar particles striking the earth – as does carbon(14).

Carbon(14) is an isotope of carbon and is used to date carbon-containing materials. That is, carbon(14) decays at a known fixed rate: it has a half-life of 5,770 years. Each 5,770 years half of the carbon(14) nuclei that remain will decay. Hence, the decay, or radioactivity, of a carbon-containing entity, such as the Dead Sea Scrolls, or a sunken Viking ship, or the remains of a caveman's campsite, determines how old it is.

The background of radiation is measured in the unit called a rem (roentgen equivalent in man). Each of us receives on an average about 0.1 rem per year, generated by ourselves and our environment. This varies from about 0.1 in New York to 0.3 rem in Denver, Colorado. Such levels of radiation damage only about one molecule in 100 million per year per person. The Atomic Energy Commission has set the safety level at about 2 rems per year per person. A fatal dose is approximately 10,000 times the total yearly radiation absorbed by a person from the natural background count, but it must be received in a single dose.

Some of the better-known radioisotopes, generated by uranium or plutonium atomic bombs and by nuclear power plants, are strontium(90), which accumulates in bone; iodine(131), which accumulates in the thyroid gland; cesium(137), which accumulates in the nervous system; and iron(59), which accumulates in the bloodstream.

Radioisotopes will accumulate wherever their nonradioactive "brothers" accumulate. Hence, radioisotopes released into the environment will not simply disperse but will become concentrated in some vital organ as the living body selectively gathers specific elements in specific tissues to fulfill specific life functions. Their residence-time in their respective tissues is the biological half-life of the element. For example, the biological half-life of iron (for all isotopes if iron, nonradioactive and radioactive) is about 60 days in red cells; this means that every 60 days the body reprocesses one-half of all the iron in its red cells. This total "travel picture" – the sequence of chemical events in the history of an isotope, the selective processes that gather it in a specific tissue, and its residence time there – are called the ecological path of an isotope. The path is specific and unique to each isotope.

Many people are rightly concerned that radiation, either from nuclear testing or from atomic power plants, will adversely affect the genetic integrity of man. They reason that the trace quantities of radioactivity added to our environment from the atomic energy program will cause mutations in our genetic makeup that will accumulate over the years and thereby ultimately annihilate our race as we know it.

But what about the greater danger to our genetic integrity from the "benefits" of modern civilization?

Most genetic mutations are harmful, considering the highly complex system of our genetic code. The effects wrought by random mutations are most likely to be detrimental to the functioning of all other integrated components. Drop a watch and see what the probability is of its keeping better time.

Furthermore, it does not matter how mutations arise, whether by radiation or by a chemical agent. Mutations are mutations.

The critical questions are: what are the major sources of mutations for man? What is the major threat to the repository of human traits, our genetic pool? What, if anything, disrupts our genetic coherence?

The natural background of radiation within us and our environment, present for eons before man evolved (about 0.1 roentgen per year per person), produces about 5 percent of all our genetic mutations; the other 95 percent arise from other natural chemical causes. The average chest X-ray delivers about 0.1 rem; this is about equal to a whole year's worth of background radiation. All the atomic testing in the atmosphere since 1945 has produced about 0.1 rem per person.

Atomic power plants will emit about one-millionth of a rem[6] per person per year, far below the background exposure of 0.1. Obviously, the radiation from a full-scale atomic power program of even 1,000 power plants will be far below even the background level of radiation.

Let's now look at the "benefits" of modern society. In prehistory, the mutation rate equaled the elimination rate of harmful mutations. An equilibrium existed between the accumulation and the elimination of defects.

Modern civilization has changed all that. Our sophisticated technology and medicine have saved lives that would normally have been eliminated. Our genetic pool is filling up with defective genes because the elimination rate has been artificially reduced.

Estimates are that if civilization is saving one-half of the defective mutations that arise spontaneously per generation, then in 8 generations (250 to 300 years) we will have accumulated the genetic defects equivalent to having been exposed to a nuclear war in which we received the same radiation as did the survivors of Hiroshima.[7]

In other words, only nuclear war can accelerate the defects in our genetic integrity to an extent equal to the devastation of the gene pool wrought by modern living standards.

Apparently, the means for the destruction of mankind may be

concealed in the ramified benefits of modern living standards, as well as in a few cubic feet of nuclear warheads.

ATOMIC ENERGY IS SAFE

Common fears concerning nuclear power are: (1) the radioactive waste problem; (2) the possibility of an explosion; (3) atomic plant-site emissions, and (4) waste heat problems.

(1) Even though large quantities of radioactive isotopes will be generated by atomic power plants, these isotopes can be safely stored underground until their radioactivity subsides to background levels. The Atomic Energy Commission estimates that about 500 to 1,000 years are sufficient to render wastes only slightly above background activity.

How safely can radioactive wastes be stored underground? On the basis of "absolute zero probability of error" there is no human activity that is completely free of risk. Of all the problems attending nuclear power, this would appear to be the least worrisome.

The storage of nuclear wastes has caused concern due to the amounts of radioactivity involved. For example, we've seen that by 1985 the electric power output of the United States may be 750 billion watts. If 15 percent of this total were to be generated from nuclear power, the radioactive wastes would be enormous. Taking the figure 1.2 tons of uranium per year per billion watts, 135,000 kilograms of uranium would be required, which will produce almost 135,000 kilograms of radioisotopes per year. Considering that the first atomic bomb was about 2 kilograms in mass and yielded 2 kilograms of radioisotopes, then the radioactivity of 67,000 atomic bombs is involved in the example above. Or, generally speaking, the quantities of radioactivity produced will be of warfare proportions, much larger than the tiny quantities needed for medical research or anything else.

In industry, the radioisotopes would amount to about 1,000 cubic feet of liquid waste per year per 1,000-megawatt power plant.[8] The wastes would be pumped underground into leak-proof salt caverns, or into caverns created by nuclear explosions for this purpose.

However, the major cause for concern is not the siting of nuclear wastes but the transportation by truck or train from the chemical processing plants to the burial site itself. The safety of this aspect cannot be foreknown.

(2) Nuclear power plants are explosion-proof; no accident or deliberate attempt on the part of a saboteur could cause a nuclear explosion. The arrangement of the fuel pellets absolutely precludes this.

A bomb requires almost pure uranium(235), whereas a power plant has only enriched uranium [3 percent uranium(235)]. This alone precludes a nuclear explosion, but also there is the critical mass and critical time requirements for an explosion. A critical mass is the necessary weight which under specific conditions is required to achieve detonation. However, it is not the absence of sufficient mass that promotes protection against a nuclear explosion but the time required to bring it together (milliseconds). Even if the fuel elements were pure U(235), in a melt-down the critical mass would be achieved far too slowly for a nuclear explosion to occur.

There is a special concern with the breeder reactor, which will use liquid sodium as a heat transfer medium. What would happen in the event of a mechanical failure that led to a loss of sodium circulation through the reactor core? In such a case, the core would overheat to about 3,000 degrees, melting the core and its steel casing, and resulting in heavy radiation to the environment. However, all breeder reactors will be equipped with a back-up cooling system to prevent this loss of core-cooling. And there is further protection in the concrete shielding of the reactor, which is designed to retain all the melted components.

In the twenty years or so that prototype and commercial nuclear plants have been in operation, accidents have caused seven deaths.[9] In 1957, a reactor failed in England, along with its safeguards. Another accident, in Idaho in 1961, resulted in the death of three staff members. A fuel melt-down occurred in the Enrico Fermi Power Plant in 1966, causing some alarm. A slight increase in radioactivity was detected within the plant immediately after the accident, but it was all contained within the plant. It required approximately four years to restore the plant to operational status.

But seven deaths due to nuclear power plant operations is an enviable record compared with deaths caused by normal activities involving other technologies. The number of people in this country killed in motor vehicle accidents from 1945 through 1968 is approximately 930,000.[10] It is impossible to estimate how many thousands of people have had their lives shortened as a consequence of fossil-fuel pollution, particularly during times of temperature inversions. Air pollution alone is implicated in the deaths of approximately 10,000 people a year in New York City.[11]

No new technology can burst suddenly onto the power scene and claim complete safety. It takes years of careful, painstaking, laborious work to construct and test prototype models, in many different configurations, so that sources of risk can be identified and corrected. It takes additional years just to develop the many technologies prerequisite to the uses of these new power sources. We need hardly be reminded of the large number of unmanned and manned space flights that preceded the Apollo mission that finally placed a man on the moon.

(3) As for emissions, nuclear power plants are less offensive than fossil-fuel plants. Atomic plants emit no sulfur dioxide, no particulates, no nitric oxides, no carbon monoxide, and even less radioactivity than conventional fossil-fuel plants on a megawatt basis. The reason is simple. No coal seam is 100 percent free of uranium or any other radioactive isotope. By burning coal, the impurities are released in the gases. Neither is any other ore or rock or deposit of any kind free of impurities. Every natural resource has impurities in it.

(4) Thermal pollution has become a forum of conflict between power companies and ecologists, even among ecologists themselves. The argument usually starts out like this: by the year 2000 perhaps one-third of all United States streams may be used to cool atomic power plants.[12] Hence the cooling water will be released a few degrees hotter than it was at its entrance into the power plant. This degree rise could nevertheless be potentially dangerous to river life. The preferred spawning temperature for most species of

fish is below that of their preferred temperature ranges in general. Hence, the heating of river water may be injurious to the critical reproduction stage of aquatic life.

The opposition argues that slight temperature rise will be mostly localized and not affect reproduction, but will, in fact, speed algae growth and thereby increase the fish's food supply. In fact, oysters are now successfully grown in the warmer effluents of some power plants.

Ultimately, the ecological consequences will depend upon the amount of heating and upon the river's normal aquatic species. Only a quantitative study on each river can decide the issue.

When all the facts are studied, the atomic energy program will be recognized as one of the safest modern technologies engaged in by man. Unfortunately, the relentless attack on nuclear power in all cases may not be just a misunderstanding of nuclear power but something worse – a prejudice against modern science.

NOTES

1. Committee on Resources and Man, National Academy of Sciences, National Research Council, *Resources and Man* (San Francisco: W. H. Freeman and Co., 1969), ch. 8.
2. "Nuclear Power Plant Safety at Issue," *Chemical and Engineering News* (January 24, 1972).
3. R. L. Faulkner, Remarks at the Conference on Nuclear Fuel-Exploration to Power Reactors, Oklahoma City, Okla., U.S. Atomic Energy Commission press release, May 23, 1968.
4. Committee on Resources and Man, *Resources and Man,* ch. 8.
5. Ibid.
6. Glenn T. Seaborg and William R. Corliss, *Man and Atom* (New York: E. P. Dutton and Co., 1971), p. 77.
7. H. J. Muller, "Radiation and Human Mutations," *Scientific American* (November, 1955).
8. The U. S. Atomic Energy Commission, *The Nuclear Industry* (Washington, D.C.: Government Printing Office, 1971), p. 158.
9. Neil Fabricant and Robert M. Hallman, *Toward a National Power Policy* (New York: George Brazillier, 1971).
10. *1972 World Almanac* (New York: The News), p. 88.
11. Thomas A. Hodgson, Jr., "Short-Term Effects of Air Pollution on Mortality in New York City," *Environmental Science and Technology* 4 (July 1970), pp. 589-598.
12. Seaborg and Corliss, *Man and Atom.*

Nuclear Fusion: An "Ultimate" Answer to Our Future Energy Needs

Ask a benched athlete (either temporarily retired or nursing an injury), "Why is it so difficult to come back?" Unquestionably he will mention timing. For the golfer, it's the timing of his swing so that maximum acceleration of the club head is achieved at the moment of impact with the ball; for the boxer, it's timing his punches to achieve maximum impact on his opponent; for the tennis buff it's the timing of his strokes to achieve proper control over the direction and speed of the ball. If the timing is off, even a great athlete will appear mediocre.

In a sense, we could make a strong case for timing as the central issue of this book. We have seen that the problems of energy delivery will reach a critical stage within the next two decades. But have we trained ourselves to meet this encounter? Synthetic fuel production should have begun a decade ago. It has yet to begin. The breeder reactor should have started two decades ago. It will not supply half of our electrical needs until after the year 2000, at the earliest. The use of solar energy is in the distant future.

What is most disturbing is that Russia does not face this timing problem. There will be no gas or oil shortage in Russia, not for many decades. If as a result of the Middle East political situation an oil shortage should develop, Russia has sufficient natural gas deposits to carry it for a very long time to come.

It's interesting to ask why Russian enthusiasm has waned in the race for space. Do its leaders recognize that the enormous expenditure of funds and resources in this venture is profligate at a time when the problems of energy resources are paramount? Clearly Russia is very deeply involved in the development of energy sources. She has discovered enormous natural gas fields in Siberia and her Volga River hydroelectric power plant is the largest in the world. She will have a 600-megawatt breeder reactor on-stream between 1973 and 1975. The United States will have a 1,000-megawatt breeder reactor on-stream in the early 1980s. Both countries presently have small-scale, noncommercial, experimental breeder reactors in operation. Neither country has developed a controlled and sustained fusion reaction.

THE NATURE OF FUSION

Nuclear power may be derived from one of two processes – fission or fusion. As we have seen, nuclear fission is the process in which a nucleus [for example, uranium(235)] splits to form nuclei of lower atomic number and lower atomic weight. In this process there is a mass loss that is transformed into energy in accordance with Einstein's formula $E = mc^2$ (energy equals mass loss times the velocity of light squared).

Nuclear fusion is a process in which atomic nuclei combine to form an atomic nucleus of higher atomic number and weight (for example, four hydrogen nuclei combine to yield helium, which is what occurs in the sun's nuclear "power plant"). In this process there is also a mass loss that appears as energy; the same equation applies, $E = mc^2$.

Not only does the fusion of hydrogen nuclei power our sun, but most of the billion of stars in the universe. This process of hydrogen-hydrogen fusion has not been achieved on earth. The "hydrogen bomb" employed the fusion of two isotopes of hydrogen, namely tritium with deuterium. Therefore, of the fusion reaction possibilities, the one that has been achieved on earth is tritium-deuterium. However, deuterium-deuterium is believed achievable; hydrogen-hydrogen is believed unachievable on earth for physical reasons.

The problem is to achieve a sustained and controlled fusion reaction; probably this will be accomplished first with tritium-deuterium and, perhaps, thereafter with deuterium-deuterium. The largest energy bank is that residing in deuterium. Deuterium is found in water; it is an isotope of hydrogen and its relative abundance in hydrogen is one part in 6,500. The deuterium in one gallon of water would yield the energy equivalent of 300 gallons of gasoline, and can be extracted for a few pennies.[1] *If all the deuterium in the oceans (200 trillion tons) were to be harnessed, the total energy released would be one hundred million times all the energy reposed in the world's initial supply of fossil fuels.*[2]

The deuterium-deuterium fusion reaction expressed schematically is:

$$5\ H(2) \rightarrow 2\ He(4) + \text{energy} + \text{neutrons}$$

Five deuterium nuclei [H(2)] fuse to form two helium nuclei [He(4)] and emit neutrons plus energy.

The fusion of deuterium with deuterium has not been achieved, but if achievable then the world's energy reserves could serve man for more than a billion years.[3] This source of power would yield no radioactive wastes (unlike uranium or plutonium). Also, there would be no breeding rate or fuel production limitations. In addition, this source of power would yield no radioactive emissions from the power plant itself.

Deuterium-deuterium fusion holds out other great prospects. This fusion process would occur at hundreds of millions of degrees centigrade. By a unique design, called the "fusion torch," reactors running on this fuel could be used to vaporize waste materials into their atomic components. This would make the total recycling of every man-made product possible. Steel, metals, old cars, garbage – anything could be vaporized. From our industrial wastes would come their constituent atoms. No radioactive wastes (radioisotopes) remain – in contrast to nuclear fission.

Fusion plants powered by the deuterium-deuterium reaction would run at very high temperatures and hence very high energy conversion efficiencies: atomic power plants (fission) are 30 percent efficient; but fusion power plants (deuterium) could be 90 percent efficient. Here is a solution to thermal pollution; the heat loss problem would be greatly lessened.[4]

Unfortunately, controlled deuterium-deuterium fusion will probably be achieved after the achievement of tritium-deuterium fusion, which is easier to effect but which has only been achieved in the hydrogen bomb and is not yet controlled.

FUSION SCHEMES: SUPERIOR TO FISSION

The fusion reaction that appears closest to being achieved is the fusion between tritium and deuterium (two isotopes of hydrogen). Whereas deuterium occurs in water (one part in 6,500) tritium does not. Tritium is radioactive and decays with a half-life of 12 years. Hence, it must be bred from another source. The source of tritium is lithium(6) – Li(6) – one of two isotopes of lithium in nature. When lithium(6) is bombarded by neutrons it is transmuted into helium and tritium. This means that the energy bank available to man from the tritium-deuterium fusion process is limited by the availability of the isotope Li(6), which is 7.42 percent of all lithium. The energy bank for tritium-deuterium fusion reaction is only about equal to that of the fossil fuels.[5]

The reaction for tritium-deuterium fusion is:

$$\text{Li}(6) + \text{H}(2) + \text{neutron} \rightarrow 2\ \text{He}(4) + \text{energy} + \text{neutrons}$$

A lithium(6) nucleus, Li(6), plus a neutron produces a tritium nucleus (not shown: it's bred and consumed in the reaction) as an intermediate which in turn reacts with deuterium, H(2), to yield helium, He(4), plus energy plus several neutrons.

A novel idea is to ignite the tritium-deuterium by a laser beam. This design calls for the fuel to be in a frozen state and then ignited by a laser beam while in the vortex of a liquid lithium blanket. However, the most promising design calls for the electronic heating of a gas to produce a plasma. This is the design that we shall discuss.

The fusion reaction begins at about 50 million degrees centigrade. Such high temperatures preclude the use of all materials for containment of the reaction: it is contained in a "magnetic bottle." At 50 million degrees deuterium and tritium atoms are so energetic that they have been stripped of their electrons; this results in a mixture of atomic nuclei and electrons called a plasma.

There are three main problems that must be solved before controlled fusion can become commercial: one is the containment of the plasma in a suitable magnetic bottle, another is the sustaining of the fusion process over extended periods of time, and the third is the conversion of the released fusion energy into electricity.

The most promising design for a tritium-deuterium reactor is known by the name of "toroidal configuration." The plasma would have the shape of a doughnut about 3 feet in diameter and one-half-foot thick. The plasma would contain tritium and deuterium nuclei at over 100 million degrees centigrade. Upon fusion of the tritium and deuterium nuclei in the magnetic bottle, neutrons would be released that carry about 80 percent of the released energy. These neutrons would escape from the magnetic bottle (because they carry no electrical charge) and penetrate a surrounding blanket of molten lithium, causing further heating. The lithium would then generate more tritium as it absorbs neutrons. Heat from the lithium blanket would drive a steam-driven turbine to generate electricity. The tritium generated in the blanket would be separated and used with additional deuterium to continue the process.

D-T FUSION POWER PLANT

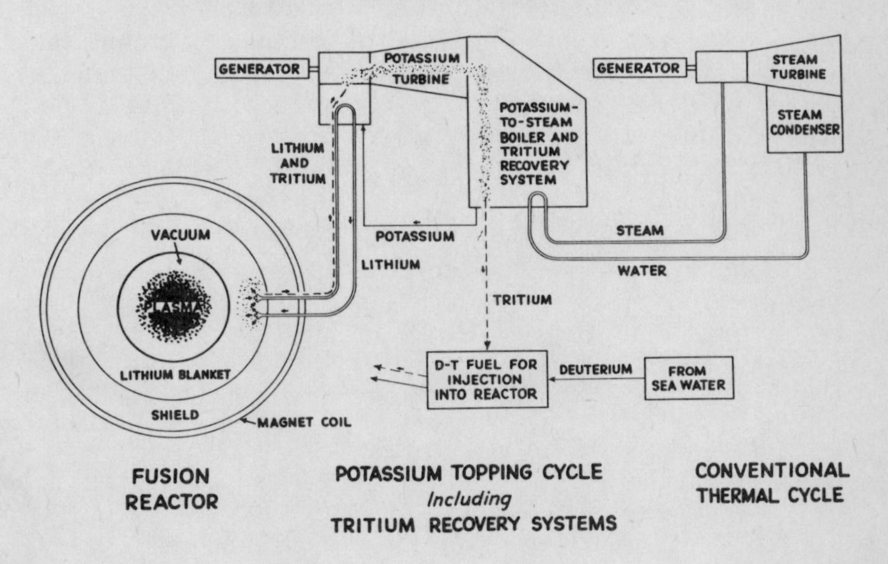

If the experiments with the toroidal (doughnut-shaped) configuration prove successful, power plants of 500-to-1,000-megawatt capacity could be built with plasma sizes of 100 feet.

This type of fusion reactor would generate electricity at temperatures comparable to a fossil-fuel plant (only the plasma is at 50 million degrees); hence its energy conversion efficiency would be only about one-third. Only the deuterium-deuterium fusion reaction offers the possibility of direct energy conversion into electricity without a heat transfer medium. Nevertheless, there are several advantages to tritium-deuterium fusion over fission (uranium or plutonium).

(a) The only radioactivity that could be emitted from a fusion power plant would come from the leakage of tritium – and this can be controlled. In fact, if America drew all its power from fusion, the tritium leakage would raise the radioactivity of the environment by only 0.1 percent of its natural level (about 0.1 rem), and there would be no radioactive wastes to be disposed of.[6] On the other hand, based on 1.2 kilograms per megawatt per year, if we drew all our power from fission we would generate about one million kilograms per year of radioactive wastes; this is equivalent to the radioactivity produced by about 500,000 atomic bombs (of 2-kilogram mass). This quantity of radioactive wastes would have to be processed each year – from fuel processing to shipping to storage sites.

(b) Tritium is similar to hydrogen in chemistry; hence, this radioactive isotope will not accumulate in the food-chain as will some of the radioactive wastes of the fission reaction. Tritium is one of the least biologically hazardous radioisotopes, whereas plutonium and some of its fission fragments [strontium (90), iodine(131), and cesium(137)] are the most biologically hazardous.

(c) Tritium must be bred from lithium(6), but its doubling time is only two months;[7] hence, there is no problem with building up a fuel inventory as there is with plutonium.

(d) A deuterium-tritium fusion power plant would be safer than a breeder reactor because there is no loss of cooling contingency: if the circulation of lithium should fail for mechanical reasons the fusion reaction would stop (not overheat as in the breeder reactor).

BUT HOW SOON?

Dr. Roy Gould, assistant director of the Atomic Energy Commission's Thermonuclear Research Division, has cautioned Congress that decisions will have to be made soon as to which types of fusion reactor we must research if the budget remains meager.[8] Since there are numerous reactor designs, each with its own special advantages, Dr. Gould believes that we should research several major reactor designs.

An all-out approach could cost a mere $1 billion through 1977. The Apollo effort cost upwards of $40 billion; support for nuclear fusion research runs about $30 million per year to date – about one percent of the Apollo cost. America spends more money in one year on deodorants (approximately $390 million) than it has on fusion research throughout the past ten years (approximately $300 million). Perhaps national prestige can be equally derived from achieving a nonpolluting and eternal power source as it is from moon walks, if one is calculating national prestige possibilities.

We are now planning to spend $1.6 billion on cancer research over the next three years and $5 billion on a space station over the next five years. Both these projects are important, but so is power procurement.

The fusion scientists believe that no breakthrough will suddenly make fusion power a reality. However, various research efforts indicate that fusion power is possible.

We have seen only some of the fusion reactor design possibilities. Fusion power, as one can imagine, is far more flexible than fission power. Furthermore, the research budget that is required is minuscule compared to our breeder reactor research budget or each of our several space programs. Surely, here is where money will be well spent. Surely, here is where scientific manpower, our real strength, will be wisely used.

However, whether it be of fission or fusion, we must not forget, the nuclear future will always require some fossil fuels, not as energy, but as a source of carbon compounds. Even the unlimited energy of fusion power will not obviate the need for carbon chemicals for the plastics industry, carbon fuels for lightweight aircraft, and carbon (coke) for the steel industry.

Whatever the future holds, the point should be clear: the fossil fuels could provide us with energy for only a few hundred years, but as chemicals they could last for a hundred thousand years in the production of plastics and other basic chemicals. As a source of coke, to produce steel, coal's lifespan would be increased hundreds of years (and here a recycling possibility exists through fusion power so that carbon for this purpose could be an inexhaustible resource). Fossil fuels for lightweight aircraft come to about 2 percent. This small consumption level will protract our reserves of fossil fuels for this purpose into centuries, and hence, give us time to develop a replacement for even this use of our finite fossil fuels.

If economics were the only factor to consider, it would be easy to recommend the crash construction of an enormous number of burner-reactor power plants to replace conventional plants that presently consume oil and gas reserves. By combining this strategy with the stepped-up development of synthetic fuels, we could appreciably reduce our dependence upon foreign sources of energy during the next two decades.

However, this is not recommended because the extensive use of uranium(235) would foreclose the possible future employment of the breeder reactor. As for the breeder reactor, we have seen it cannot possibly come to our rescue during the next few decades. The ten-year time for doubling the fuel inventory places a physical restraint on the phase-in time, over which we have no control. On the other hand, if we hoard our reserves of U(235) until the breeder reactor is on-stream, this latter power source would serve us for many years to come. But its possible future use may be strictly academic if our industrial life-support systems have collapsed before the breeder reactor is ready.

Some experts believe an all-out research effort could establish the feasibility of controlled fusion in five years.[9] If a controlled reaction is achieved, fusion power could become a reality before the breeder reactor is providing significant amounts of power since there is no fuel inventory problem. Thus, the breeder reactor could become obsolete during its early stages of use.

If we were willing to take an enormous gamble, we could bank on the development of controlled nuclear fusion, scrap the breeder reactor program, and adopt the strategy mentioned above – drastically expand our burner-reactor program to meet the energy crisis. In view of the uncertainty of fusion, however, even with an all-out research effort, it is unwise.

Nevertheless, the feasibility of fusion power is the key to determining which energy option we should pursue. It is for this reason that we need a research effort on the scale of the Manhattan Project (the development of the atomic bomb) to determine the feasibility of fusion. The financial risk is small when contrasted to the possible benefits – cheap, relatively nonpolluting, efficient, long-lasting, and flexible power.

UNITED STATES POWER
PRODUCTION POSSIBILITIES

ETERNAL SOURCE	POWER POTENTIAL (%)*	EFFECT ON ENVIRONMENT
Hydroelectric	8	no pollution
Tidal	1	no pollution
Geothermal	1	negligible pollution
Solar	1,000 plus	no pollution
NONRENEWABLE SOURCE	**PROBABLE LIFESPAN**	**EFFECT ON ENVIRONMENT**
Oil and gas	2 to 3 decades	heavy pollution
Coal	200 years (?)	heavy pollution
Nuclear fission	100 to 1,000 years	radioactive waste problem
Nuclear fusion	infinite	no pollution

*Compared to the present United States power expenditure (2 trillion watts).

Before we leave the subject it is important to restate that fusion power, if achieved, will deliver its energy in the form of electricity. Hence, to employ fusion power we need the capability of deploying electricity, or the capability of using a secondary fuel created with the aid of fusion power. Such a secondary fuel could be hydrogen gas created from water.

We will see in the next chapter that our nation, and the world, is unprepared to deploy electricity due to a very severe shortage of copper (and aluminum) and the creation of hydrogen gas awaits the development and deployment of a substantial power source, such as the breeder reactor, or fusion power, or solar power. We shall return to this topic in a later chapter. The deployment of fusion power will prove to be as complex a problem as the creation of a workable fusion reactor itself.

NOTES

1. Glenn T. Seaborg and William R. Corliss, *Man and Atom* (New York: E. P. Dutton and Co., 1971).
2. Committee on Resources and Man, National Academy of Sciences, National Research Council, *Resources and Man* (San Francisco: W. H. Freeman and Co., 1969), ch. 8.
3. Ibid.
4. Thomas E. Feare, "Fusion Scientists Optimistic Over Progress," *Chemical and Engineering News* (December 20, 1971), pp. 29-38.
5. Committee on Resources and Man, *Resources and Man,* ch. 8.
6. Lawrence M. Lidsky, "The Quest for Fusion Power," *Technology Review* (January, 1972).
7. William E. Cough and Bernard Eastland, "The Prospects of Fusion Power," *Scientific American* (February, 1971), p. 50.
8. Feare, *Chemical and Engineering News,* pp. 29-38.
9. Dr. Roy Gould, assistant director of the AEC, in Feare, ibid.

The Electrical Requirements of a Nuclear Age: We're Unprepared

We have seen that the industrial power base in the future must be either nuclear or solar. Whether we can draw our energy from fission (uranium or thorium), or fusion (deuterium or deuterium-tritium), or solar power, we shall only be able to harness these energy sources as electricity. Electricity will be the major mode of power expression in the future, to be used directly for heating, transportation, communication, and manufacture or transformed into a secondary fuel such as hydrogen.

THE LOOMING COPPER SHORTAGE

The electrical requirements of the future will be enormous. What will carry the current? Copper's exhaustion date appears to be around the turn of this century.

During the last 40 years in the United States, aluminum production has increased over 100-fold; copper production over tenfold; steel production over tenfold; plastics production over 500-fold; and so on; lulling us into an assurance that natural resources are limitless because newly discovered scientific principles can engineer a resource out of almost anything, as all the great precedents in industrial history would seem to imply. But is this the case?

Is there no end? Is any grade of ore workable? On a laboratory scale, copper can be extracted from the poorest ore. But on an industrial scale, if the energy required by the equipment used to acquire copper metal from its ore were to reach disproportionate levels, then copper production would become too disruptive of the total economy.

Nor are there unlimited substitutes. If copper runs out, aluminum might be used as its substitute. If aluminum runs out, then perhaps in the future, conducting plastics will be invented to substitute for aluminum. But plastics are at present just a dream.

Copper's chief value lies in its ability to conduct electricity. The only other metals that are as good a conductor are silver and gold, both too rare to be used. Aluminum is second to copper as a conductor, but its melting point is lower, 660° to copper's 1083° centigrade.

How much copper is there? The United States possesses about one-fourth of all known copper deposits. The known deposits amount to approximately half of all the estimated copper on earth that is recoverable. Our present reserves are estimated at 50 million tons; the world's reserves at 200 million tons.[1]

By comparing our reserves to our consumption rates we can see that a copper shortage is imminent. Since copper consumption fluctuates each year due to stockpiling and production contracts, we'll take the latest five years' average for production and consumption.[2] We are presently consuming over 2 million tons of copper per year. Of this, about 1 million tons come from the smelting of domestic ores, about 0.6 million tons come from recycled copper (about 40 percent of domestic production) and approximately 0.4 million tons are imported (80 percent of this from Chile).[3,4]

At the rate we are consuming domestic reserves – 1 million tons per year – our 50 million-ton reserves would last for only 50 years. But our growth rate in consumption of copper is 3.3 percent per year compounded.[5] This accelerated consumption rate places copper's lifespan at about 30 years for domestic reserves.

The conclusion is clear: at the turn of the century we hope to be on the threshold of a nuclear age. The nuclear age will demand prodigious increases in the deployment of electricity. But the key to this deployment, copper, will be near exhaustion at the very time when the need for it is at its zenith.

Finding new deposits of copper will become progressively more difficult. From the Civil War to 1940 the United States mined and used copper ores of about 3 to 1.5 percent copper by weight. However, for the last 20 years we have been obliged to use ores of 1 percent or less of copper. The best copper deposits have been found and exploited. Now we are searching harder for poorer grades of ores.

Since World War II the United States has been an importer of copper. The percentage of copper imported fluctuates widely, but it averages over the last 5 years to about 17 percent of our total needs. The United States possesses the largest share of world reserves (28 percent), followed by Chile (19 percent), and Zambia and the U.S.S.R. just behind Chile.[6]

NUCLEAR STIMULATION MINING FOR COPPER

Are there other ways of obtaining copper?

Kennecott Copper Company and the Atomic Energy Commission are presently discussing plans for Project Sloop.[7] This project would explode a 50-kiloton atomic bomb underground in a copper-bearing ore. Approximately one million tons of rock would be crushed in the cavity created by the explosion and the copper content of the ore would be extracted with sulfuric acid pumped into the cavity. Project Sloop is a concept, not an actual undertaking.

With most ores of about one percent copper at best, this means that no more than about 10,000 tons of copper could be obtained per nuclear blast. Since our country's annual consumption of copper is over one million tons, we would need at least one hundred nuclear explosions per year to satisfy our copper needs by this technique.

Over one hundred nuclear explosions per year to stave off copper starvation by 2000 is quite a grim prospect, considering the safety factor alone.

THE MYTH ABOUT SEA-WATER COPPER

The world's oceans possess about twenty times (about 5,000 million tons) the copper estimated on land – but is it obtainable?[8]

The copper in the oceans is very dispersed. There are only about 14 tons of copper per cubic mile of sea water.[9] At our annual needs of about 1 million tons of domestic copper we would have to process about 70,000 cubic miles of sea water to get this amount of copper. Imagine the equipment and energy requirements to process 330 trillion tons of sea water! This is about 330,000 times the amount of oil we process per year.

It can't be done – even with fusion power. The dispersion problem of copper in sea water (like the dispersion of solar power) is a sublime barrier.

At present we have about 150 kilograms (330 pounds) of copper per person in the industrial environment of the United States. This amounts to about 30 million tons of copper. The largest share is used in the various electrical industries. Our present electrical power is 375 billion watts out of a total economy of 2,000 billion watts (2 trillion), or 19 percent. As our deployment of energy in the form of electricity rises to ever higher percentages of an increasing total power need, we will require more copper. The present copper growth rate of 3.3 percent per year cannot remain constant – our energy growth rate is 7 percent per year, and the future will call for a greater proportion of electricity, not less. Hence, we shall unquestionably witness an upsurge in copper needs as we approach the year 2000.

A major question that will affect our economy is the degree to which nuclear energy (fission or fusion) or solar power would be deployed directly as electricity or as a secondary fuel. We've mentioned atomic energy, as electricity, being used to cleave water into its constituent atoms, hydrogen and oxygen. The hydrogen gas could be piped around the country and used just as is natural gas. Hydrogen could be used as a substitute for gasoline in a car, for oil in a heating plant, and for gas in a power plant that generates electricity. The use of hydrogen gas could greatly reduce the electrical requirements of a nuclear age. But by how much? Again, only the engineering profession could give an accurate estimate. Even so, the electrical requirements of a nuclear age will be much greater than today. An atomic future and an electrical future are basically inseparable.

IS ALUMINUM THE ANSWER?

Aluminum is at present used in some high-voltage transmission lines. Can it completely substitute for copper? Or significantly? These questions can be answered only by the electrical engineering profession; they should address themselves to this important question while we still have the time to prepare for the massive changes that will be wrought in all our life-support systems by changing the conductor of electricity.

Estimates of world reserves of aluminum are very approximately 1,000 million tons.[10] This is only about five times the reserves of copper. The electrical needs of the future (the atomic age), we have seen, will easily make aluminum just as short-lived as copper is at present.

Further complicating the picture is the fact that aluminum is obtained by electrolyzing a molten mixture of ore (bauxite). Thus, the production of aluminum itself requires large amounts of electricity.

The United States is in an especially precarious position with respect to aluminum. We must import 83 percent of our needs (5 million tons per year).[11] The nations with the highest percentages of world reserves are Australia (33 percent) and Guinea (20 percent). Russia, with a lower aluminum consumption rate, is at present self-sufficient.[12]

HELIUM'S UNIQUE ROLE IN SUPER-CONDUCTIVITY, BREEDING, AND FUSION

One way to conserve on copper is to use a tin-niobium alloy for a transmission cable. These cables would connect a power plant with a city and be run underground. However, the tin-niobium alloy conducts well only at low temperatures; in fact, it becomes a superconductor near absolute zero (-459° Fahrenheit, as distinguished from zero). At close to absolute zero it conducts electricity with almost no loss of power. There are about 20 elements (metals) or combinations of their alloys that become superconductors near absolute zero (-459° F.). However, at higher temperatures (20° above absolute zero or higher), they become very poor conductors of electricity and are not suitable for electrical purposes. To keep a power cable refrigerated to near minus 459 degrees would require a refrigerating system placed along the path of the underground cable; these systems would operate on helium gas as a refrigerant. No other substance can substitute for helium as a refrigerant; its physical properties are unique.

This proposal would not save much copper or energy. The superconducting cable would be used only for large current-carrying cables, not for most electrical devices; and power loss in transmission of electricity is about ten percent of total electric power, reducing the value of the saving through superconducting cables.

Another problem exists with helium: it is in short supply. The National Academy of Sciences estimates that by the year 2000 we will be running short of helium. Helium has unique properties that cannot be duplicated by any other material as far as cryogenics (low-temperature physics) is concerned.

At present, the United States has a helium conservation program that stores about 4 billion cubic feet per year (mostly at Amarillo, Texas). Helium is obtained from natural gas, where it occurs to about 4 percent in the highest concentrations. Our helium conservation program, which would become progressively smaller as we exhaust natural gas, could store up to 63 billion cubic feet by the time we completely exhaust all the natural gas (the year 2000).[13] At the commercial rate of use of five billion cubic feet per year one can see that our helium reserves are inadequate for our proposed uses of helium.

Another tragedy of our inadequate helium reserves relates to a special kind of breeder reactor – the gas-cooled breeder reactor. This design calls for helium gas as the heat transfer medium. Its advantages over the liquid-metal breeder reactor would be that it could run at higher temperatures and hence higher efficiencies.

Helium is also a critical part of our fusion research programs. Fusion requires that a plasma of deuterium, or tritium-deuterium, be confined in a magnetic bottle. Strong magnetic fields are, therefore, required. The strongest magnetic fields are produced when current flows in a copper coil that is chilled to near absolute zero. It is a fascinating thought that the fusion reaction of over 100 million degrees, comparable to the sun's temperature, will be contained by magnetic walls emanating from electromagnets chilled to near minus 459 degrees.

How much helium would a full-scale fusion reactor require for its cryogenic magnets? How much helium would be recoverable from the by-products of the fusion reaction itself? If we had a two-trillion-watt economy based on fusion, the amount of helium generated would be about 250 tons per year. This is based on multiplying 2 trillion watts by the number of seconds in a year, 31.5 million, and the 240-billion joules of energy released per gram of helium formed.[14] But is it recoverable? Under the conditions at which it is produced – over 100 million degrees Fahrenheit and only 1/10,000 of an atmosphere of pressure (almost a pure vacuum) – it is doubtful that a great percentage of this helium is recoverable. In addition, the United States presently produces 5,000 million cubic feet per year of helium from gas wells for commercial consumption. Compared with the 250 tons per year above, which is only 42 million cubic feet, or 0.9 percent, we are, indeed, faced with a helium shortage of disastrous proportions.

FACING THE HARD REALITIES

We should raise some questions at this point:

(1) What is the balance between the use of a secondary fuel (hydrogen) and the direct electrical requirements of a nuclear age?

(2) What will happen to the state of our national security and balance of trade if we must import even greater amounts of copper (or aluminum) for electrical needs?

(3) What will happen to the competition between Japan, the European Economic Community, and the United States for copper? Are we on a collision course over the copper resources in Chile and Zambia?

(4) Why are the Chinese (People's Republic of China) building a railroad in Zambia toward the east coast of Africa? Will Zambia turn Marxist as did Chile?

The world is clearly experiencing a struggle for oil in the Middle East: this fact reveals the importance of energy now. Will we also see a struggle for copper (and aluminum) as the key agent to deploy electrical power?

The shortage of copper, the difficulties with substitutes, the impracticality of nuclear stimulation of low-grade ores, all plead for a national energy center that will include the study of the problems of a suitable electricity conductor for the coming nuclear age. Later we will discuss more fully the need for an energy center to consider and resolve our entire power policy before it's too late.

NOTES

1. Committee on Resources and Man, National Academy of Sciences, National Research Council, *Resources and Man* (San Francisco: W. H. Freeman and Co., 1969), ch. 6.
2. Ibid.
3. Ibid.
4. Charles F. Park, Jr., *Affluence in Jeopardy* (San Francisco: Freeman, Cooper and Co., 1968), p. 288.
5. Committee on Resources and Man, *Resources and Man,* ch. 6.
6. Donella H. Meadows, Dennis L. Meadows, Jorgen Randers, and William H. Behrens III, *The Limits to Growth* (A Potomac Associates Book. New York: Universe Books, 1972), p. 56.
7. Glenn T. Seaborg and William R. Corliss, *Man and Atom* (New York: E. P. Dutton and Co., 1971).
8. Based on 14 tons of copper per cubic mile of sea water and an ocean volume of 350 million cubic miles. *The Handbook of Physics and Chemistry* (Cleveland: Chemical Rubber Co., 1971).
9. Ibid.
10. Meadows, et. al., *The Limits to Growth,* p. 56.
11. Peter G. Peterson, "The United States in the Changing World Economy," (Washington, D.C.: Government Printing Office, 1971), vol. 2, p. 55.
12. Meadows, et. al., *The Limits to Growth.*
13. U.S. Department of the Interior, Washington, D.C.
14. Committee on Resources and Man, *Resources and Man,* ch. 8, p. 230.

Population-Food-Energy

By projecting the present population growth trend, and its consequences, we shall see that the natural resources that are required to sustain us or fulfill our future plans are simply unavailable. We must reappraise our plans, or reduce our numbers, or both. Specifically, the population and the land area needed to feed that population are on a collision course that will meet around the year 2020 (the energy shortage that is presently developing will preempt that disaster long before it has a chance to develop).

At the rate population and energy consumption are increasing, our expectations for power deployment for the year 2000 and beyond are absurd. Let us examine the basic trends that are presently undermining our power position.

The most influential of the interlocked variables are (1) population, (2) nutritional requirements per capita, (3) energy consumption per capita, and (4) materials consumption per capita. Devising systems that attempt to exploit our industrial and agricultural capacities to their limits will not produce a climate in which everyone can find his rightful chance at personal expression.

WHAT IS THE POPULATION "LIMIT"?

As we shall see, the land can out-supply the sea by at least twenty to one in terms of world food production. Hence, the "limit" of population can be assessed by considering the area and the productivity of available land.

ESTIMATES BASED ON LAND AREA[1]

- 32 billion acres is the total world land area.
- 16 billion acres of the 32 are mountains or deserts and are unavailable for agriculture.
- 8 billion acres are potentially arable; about 50 percent are now farmed.
- 8 billion acres are potentially grazeable; about 50 percent are now grazed.

From these data, the total available land that could be employed in agriculture is only about two times the present land area now in use. A good diet (2,000 kilocalories and a minimum of 40 grams – about 1.28 ounces – of high-quality protein per day) requires about one acre of arable land and one acre of grazeable land per person – or their combined totals of 16 billion acres divided by two.[2] That gives us 8 billion people who could be supported at good nutritional levels. These nutritional levels could clearly be met by the present 8 billion acres now in use for all 3.5 billion people now living, if production were evenly distributed.

An estimate of the upper limit to population should exceed 8 billion people supported by 2 acres per person because in the future we will probably see better fertilizers, storage techniques, and farm machinery; also better breeding of crops and animals. No one can foresee the exact increase in productivity per acre, but surely the farmable and grazeable land of the world can support more than 8 billion people. With just modest advances in agricultural productivity, a conservative figure would be 10 billion as the upper limit to world population – approachable with progressively increasing difficulty as land occupancy approaches 100 percent, but nevertheless a possibility.

For our estimate of 10 billion we have discounted the area required by people and their homes and factories, which is a small fraction of the total land area (70 percent of all Americans live on 1 percent of the land area).

Slow, but steady progress toward "super plants" and "super animals" will result in smaller percentages of land that must be devoted to agriculture, but the ultimate reduction in land requirements cannot be foretold. However, super plants (as the newer strains of corn) are proving very vulnerable to diseases; the "green revolution" could be undermining the genetic integrity of traditional crop species.[3]

PROPORTION OF LAND UNDER CULTIVATION[4]

REGION	PERCENTAGE OF ARABLE LAND NOW CULTIVATED
Europe	88
Asia	83
U.S.S.R.	64
North America	51
Africa	22 (due to desert and tropical limitations)
South America	11 (due to tropical limitations)
Australasia	2 (due to desert limitations)

If the oceans could supply us with food needs, all the land area could be turned over to people, cities, and so on, if we desired.

Food would no longer be the limiting factor for land occupancy. But the sea is a poor source of food. Why?

THE LAND—NOT THE SEA—DETERMINES POPULATION LIMITS

The sea, so often romanticized as a limitless source of food for mankind, is in reality insufficient for all but a fraction of our present nutritional needs. Note in the table below that primary production by green plants appears about equal to that produced on land but that the obtainable food is another story.[5] All figures are in millions of tons per year.

	ANNUAL PRODUCTION	OBTAINABLE FOOD
Land	100,000	6,000
Sea	130,000	320

How does one arrive at the conclusion that the land can outsupply the sea by twenty to one (6,000 to 320 in the chart above)? By calculating the obtainable food of the sea, not the sea's *total* life-support capacity (the weight of all living things in the sea, the biomass). The key is the life-support capacity of the sea *at a fishable level* – the total withdrawable biomass theoretically available to man that satisfies the conditions of being obtainable, non-poisonous, and self-regenerative.

By far, the greatest productivity of life in the seas takes place in the waters over the continental shelves, which constitute only 15 percent of the ocean area and about 3 percent of its total volume. Life requires sunlight and nutrients. However, only the top surface waters of the ocean are sunlit – 99 percent of the sunlight is absorbed within a depth of about 100 feet, or only the top 0.6 percent of the open ocean's 3-mile depth is sunlit. The nutrients of phosphates, nitrates, and iron are primarily found in up-wellings over the continental shelves. Hence, although the ocean is vast in volume, it is mostly barren of life except for the relatively small volumes of water over the continental shelves.

The present catch of fish is approximately 60 million tons per year worldwide.[6] This catch translates into only one-fourth of man's minimal protein needs for the present population of 3.5 billion and only 2 percent of man's food-energy needs (fish, high in protein, is low in calories).[7]

Estimates of 320 million cover all the usually eaten fish (tuna, sardines, herring, cod, etc.). This turnover remains constant over the years as fish are born and fish die. Not all 320 million tons per year can be caught; if they could, we could draw 320 million tons per year from the oceans indefinitely, as interest can be drawn from a bank account. But the oceans are too vast to be so completely fished. If we could catch more than 320 million tons per year, the fish populations would decline.

A 50-percent catch of the sea's potential yield is held to be the upper practical limit to fishing.[8]

But the world population is rising, and the sea's productivity is not. Clearly, the sea will prove to be, at best, a supplementary source of protein, and a poor source of food energy.

PLANKTONBURGERS?–IT COSTS TOO MUCH ENERGY

Could we filter the oceans in a manner similar to farming the land as some of our science writers and freethinkers propose? Sea plankton, taken unselectively from the oceans, would prove poisonous to man because of a high fluoride and silicon content. The removal of these elements is technologically not feasible by any method now known. The chemistry that could remove these poisons would necessarily destroy the nutrient value of plankton. Of greater importance is that plankton grow to the limit of about two hundredths of an ounce per ton of surface sea water, and hence, are too dilute to obtain economically. What we have is the dispersal problem again, a parallel between low plankton density and low solar energy density. Both are potential sources of energy – sunlight for industrial consumption and plankton for food consumption. Both are in "extent" sufficient; both are in "dispersal" inaccessible. For example, to get one pound of plankton would require filtering about 1,000 tons of sea water – an absolutely unreasonable energy requirement for obtaining food.

Can we irrigate the deserts to grow food? Technically, yes; practically, no. The energy expenditure to irrigate the deserts would be high and out of all proportion to the energy banks available to us, even to the rate at which we can possibly develop facilities to irrigate the deserts.

Suppose we desalinate the seas so as to irrigate enough desert to provide food for an additional one billion people (producible in about 12 years at the present rate of population growth of 2 percent per year worldwide). This would require about 1.8 billion tons of plant life annually.

Plants absorb water through their roots and pass it through their leaves. This is called transpiration. For about every 100 pounds of water transpired, the average crop gains one pound of growth. Hence, to grow 1.8 billion tons of crops would require 180 billion tons of water. To obtain 180 billion tons of fresh water per year from sea water by distillation would about *double* the world's present total power expenditure. With efficient distillation plants this figure could be reduced to 50 percent of world power output. Even so, the energy cost is prohibitive.

We can afford the high energy expenditures only for the irrigation of deserts on a small scale and for very local regions with local problems.

POPULATION AND ENERGY

Why are population projections more of an art than a science? The answer lies in the historical fact that the rate of population growth is a variable, and always has been. It changes with the conditions of life – peace or war, economic boom or bust, the subtleties of confidence in the future or in moral perspective on proper family size.

These different growth rates are hypothesized in the graph below. The plot gives the projected population of the United States by the year 2000, based on a constant rate of growth for each year of projection.

The curve can be read as follows: If the rate of growth in 1920 had held constant, then by the year 2000 the United States population would be 270 million; if the rate of growth during 1935 (the Great Depression) had remained constant, then by 2000 the population would be 160 million, and so forth. The 1965 dip in the curve shows that at present our population growth rate is declining, perhaps a consequence of moral or economic influences, or the Pill. Whatever the causes, the curve demonstrates that the rate of population growth is a variable. Hence, equations that predict future population sizes are only approximations of the real forces at work.

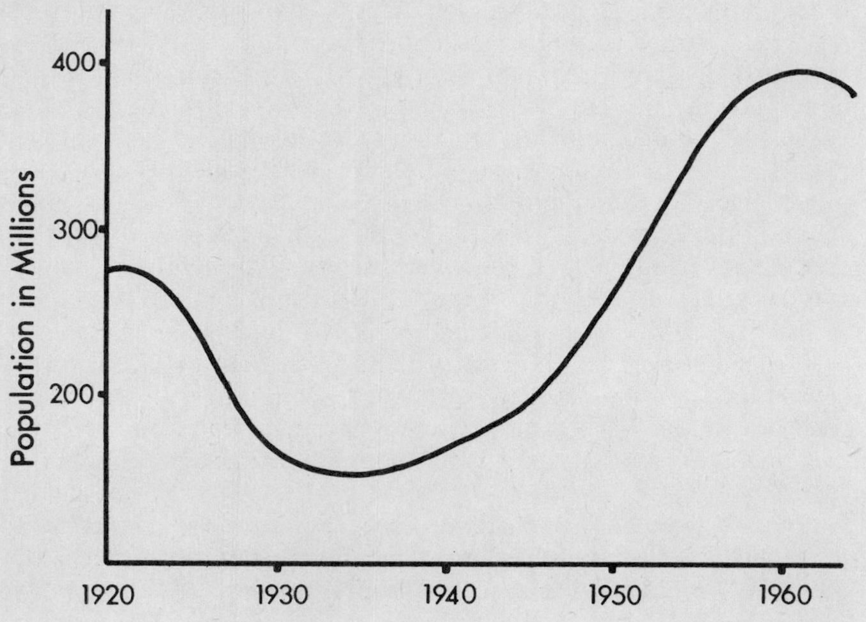

If we assume that the limit to world population is 10 billion, then at 2 percent compound growth per year, we will reach it in about 50 years, or by the year 2020.[10] Of course this limit, as we have seen, is only a gross estimate of the population that world agriculture could support at good nutritional levels.

Of greater significance than speculation on a so-called population limit is, perhaps, the recognition that present incipient stresses and consequences of increasing population will be encountered later in ever-greater magnitude as the limit is approached. And the ultimate limit appears to be only fifty years away!

At the limit of 10 billion people, the food supply would be adequate, but the living conditions stringent. All available land would be used for agriculture, for grazing, or for living sites – only the mountains and deserts would be undeveloped. At three times the present population of 3.5 billion, there may very well be endless suburbs constituting the megalopolises of the future. Certainly, man would be creating an intensively managed agriculture, a no-margin-for-error agriculture pushed to its limits. Any default by nature or error by man could result in starvation by the tens, or hundreds, of millions.

A steady progression toward an intensively managed agriculture would be required for mere nutrition. We would need governmental agencies that could run our life-support systems efficiently, agencies that could supersede our traditional free-enterprise system wherever our physical survival was threatened. Preoccupation with survival would be the dreary theme of such an age.

One can conjecture that dictatorship is ever more likely in a world where there is a monumental need for tight organization of agriculture-industry-population interactions. Survival in the future with our human spirit intact will require much more than concern over mere physical survival – the population limit should not be tested.

But the worst prospect, indeed specter, of rising population is that it would hasten an already impending power disaster because the energy required to maintain a modest living standard is simply unavailable. Consider the following energy expectations for the future.

(1) The world population could reach 7 billion by the year 2000, at the present growth rate. Suppose that all people desire a living standard measured by 3,000 watts per person; this would be twice the 1970 world average. Suppose that one-half of all energy were to come from uranium. On the basis of earlier calculations, the world would need 12,000 tons per year of fissionable uranium or plutonium. But this amount of uranium(235) is unavailable, and the breeding rate of plutonium is such that 12,000 tons of plutonium per year can't be bred until after 2050 at the theoretical best.

(2) The United States will have at present growth rates 300 million people by the year 2020. Suppose that we desire a doubled living standard (20,000 watts per person). Suppose that half of this power were to come from the breeder reactor. We would need 7,200 tons of plutonium per year. But we've seen how at the present rate of breeder reactor development we might only achieve that tonnage by 2060.

(3) With the breeder reactor not ready when needed, we discussed how we might turn to coal. Suppose 8 trillion watts were drawn from coal in the form of one-third used directly, one-third used to produce electricity, and one-third used to produce synthetic fuels. We would need to produce 18 billion tons of coal per year (30 times the 1970 rate).

WATER – A CRISIS BY THE YEAR 2000

As the population grows, naturally so does water consumption. More water is consumed by industry to produce the items that we need; water used on our behalf even though not in the home. These two phenomena give rise to an exponential growth rate for water consumption that, if projected for the next 50 years, leads to the inescapable conclusion that, by the year 2020, the United States will have a water deficit of about 30 percent.[11]

The usable fresh water runoff in the United States is that quantity that falls on watershed areas and needs little purification before it is fit for human consumption. This is about 800 billion gallons per day. The total fresh water runoff is about two to three times this amount, but it does not fall on watershed areas.[12]

UNITED STATES WATER CONSUMPTION[13]

YEAR	WATER SUPPLY
1970	60 percent surplus
2000	no surplus
2020	hypothetical 30 percent deficit

Of course, before we reach that thirty percent deficit, we'll arrive at the ceiling of available water (by 2000), after which we would be forced to hold steady on our water usage.

What can we do? Rainfall cannot be increased. The alternatives are to purify used water and recycle it, or desalt the oceans.

Logically, water recycling will no doubt be phased in as this century runs out. The most likely areas to start with are industrial plants where the contaminants are known and the water is not intended for subsequent human consumption.

Sea water distillation is energetically expensive. To double the United States supply of usable fresh water could require about 10 times the total energy expenditure (2 trillion watts) in the nation at present! On that basis, to increase our water supply by only *10 percent,* we would have to increase our power output by *100 percent.* This tenfold figure can be decreased with improved distillation efficiency. However, the heat of vaporization of water, which is constant, limits the range of reductions.

Two major techniques are under extensive research today to circumvent the high heat of evaporation of water in the distillation process. These techniques are freezing and reverse osmosis, and both have resulted in prototype desalinization plants.

The energy required to freeze pure water out of salt water is about one-seventh the energy needed for boiling. A plant at Ipswich, England, is operating commercially on this principle. But even at a six-seventh power saving, doubling our water supply would require at least double – or 4 trillion watts – our present output.

Reverse osmosis employs a semi-permeable membrane that permits water, but not salt dissolved in it, to diffuse through. Not all salts and organic matter are removable to the same degree. Hence, studies must be made to determine if this method can be applied to sea water for the creation of drinking water, or whether it is best employed in particular industrial cases where only industrially pure water is desired.

An alternative to distillation may prove to be a combination of techniques, such as reverse osmosis, freezing, and then a further purifying step to ensure sterile water. Even so, *the energy requirements to double our water supply remain prohibitively high.*

Some thinkers believe that the waste heat from power plants, either atomic or fossil fuel, could be used to distill sea water and thereby supply our fresh-water needs.

Whether the power plant is atomic or fossil fuel, we've seen that the efficiency of conversion is approximately the same – 30 percent for atomic power plants, 40 percent for fossil-fuel plants. If all of the waste heat could be employed for desalination purposes, with its known low efficiencies because of its low temperature, we would be able to distill only a tiny percentage of our fresh water at the very best.

The concept of siting atomic power plants several miles offshore to solve the thermal pollution problem will not make an appreciable dent in the onrushing water shortage problem. Even the waste heat from 1,000 atomic power plants, each of a 1,000-megawatt capacity, could contribute only a fraction of our water needs.

There is only one sensible strategy: we should plan to live within our water budget.

NOTES

1. Committee on Resources and Man, National Academy of Sciences, National Research Council, *Resources and Man* (San Francisco: W. H. Freeman and Co., 1969), ch. 3.
2. Ibid.
3. "New Food Problems Face the World," *Chemical and Engineering News* (September 27, 1971), p. 41.
4. Adapted from: Committee on Resources and Man, *Resources and Man,* ch. 4.
5. Ibid., ch. 5.
6. Ibid.
7. Ibid.
8. Ibid.
9. Ibid., ch. 3.
10. Ibid.
11. *Environmental Science and Technology* 4 (October, 1970), p. 812.
12. Ibid.
13. Ibid.

Pollution: Fossil Fuels vs. Nuclear Fuels

It is our thesis that the fossil fuels must be employed now to satisfy our immediate power needs to tide us over into the nuclear future. It is also a fact that fossil-fuel combustion is the prime cause of pollution. The cost of air pollution to the nation is estimated at about $16 billion per year, including the lowering of property values, damage to crops, serious illness, and death. Pollution, we shall see, may also cause a drastic climate change in this century and a catastrophic sea level change by the year 3000.[1]

How do we reconcile this paradox? By using coal to serve our power base and by rushing to develop the alternative power sources (fission and fusion) that could rid us of our dependency upon fossil-fuel combustion.

If the argument for the future, for some, lacks immediacy, let us look at the ravages of pollution, as much a part of us today as the air we breathe, and see why we must act soon on the power front.

THE RAVAGES OF POLLUTION FROM FOSSIL FUELS

From power plants, factories, and automobiles accrue combustion by-products called pollutants: carbon monoxide, nitric oxide, peroxides, sulfur dioxide, and particulates.

Carbon monoxide combines with iron in hemoglobin and, thereby, destroys the red cell's ability to carry oxygen. Its principal source is from our modes of transportation.

Nitric oxide by a chemical process absorbs sunlight, and thereby contributes to the oxidation of hydrocarbons into poisonous peroxides.

Hydrocarbons are the unburned molecules of a fuel and are expelled into the air from the incomplete combustion of fossil fuels, whether from a car, a home heating system, or power plant. Some nitric oxide is formed in the combustion process as an inescapable consequence of the combustion physics itself.

As we have seen, no combustion process is complete, either in an internal combustion engine, a jet engine, or a steam engine. All combustion results in the formation of particulates. Particulates are small dust particles that are individually invisible to the unaided eye. So-called "clear air" contains about 1,000 to 10,000 particulates per cubic inch. These are derived from the wind acting on the sea – to yield salt particles – or acting on the land to yield dust. The combustion of fossil fuels adds to this background level. Particulates have a very large surface-area-to-volume ratio; therefore they absorb other pollutants onto their surfaces. The background level of nature's particulates is harmless; in fact it aids in cloud cover formation for the earth. Man's contribution of particles are of another type and are toxic.

All fossil fuels contain sulfur, since sulfur is a constituent of the protein molecules of all plant or animal life; hence oil and coal (but not gas) contain about three percent sulfur. Upon combustion, the sulfur is converted into sulfur dioxide. Sulfur dioxide obstructs the body's defense mechanisms against inhalation of other pollutants and is itself poisonous.

PRINCIPAL SOURCES OF POLLUTANTS[2]

	TOTAL POLLUTANTS (Millions of Tons per Year)
Transportation	90
Power and heating plants	40
Industry	30
Garbage incineration	10

Recent discoveries show that sulfur dioxide and particulates operate in concert upon the human system to produce biological and biochemical stresses. This affects a large number of cells that protrude hair-like structures, called cilia, which are found in the throat. Under normal conditions, the cilia beat with a motion that propels mucus and waste material toward the mouth or stomach. Thus, many potentially dangerous agents never get beyond these sentries guarding the respiratory tract. However, sulfur dioxide paralyzes these cilia and prevents them from doing their assigned task. As a result, dust particles and particulates carrying pollutants adhering to their surfaces pass relatively freely through the throat and nasal passage directly into contact with lung tissue.

Theoretically, then, we could expect a relationship between air pollution levels and respiratory ailments; also more widespread effects of pollution upon bodily functioning in general. Let us look at some of the evidence.

A large number of studies have attempted to determine whether air pollution is a significant factor in the morbidity (illnesses of all kinds) and mortality rates or urban dwellers (where pollution is high) versus rural dwellers (where pollution is low). In these studies, care was generally taken to match samples of other factors, e.g., age, smoking habits, occupation, socioeconomic class. Some of the findings illustrate the high price we must pay for our continued use of fossil fuels to generate power.

(1) The death rate attributed to bronchitis is substantially higher (perhaps fifty percent or more) in "dirty cities than it is in clean country areas."[3]

(2) The incidence of lung cancer is much higher for the city dweller than it is for his country cousin. This relationship holds for both smokers and nonsmokers. In one study, the incidence of death due to lung cancer was ten times as great for the city dweller.[4] In another study, the urban rate was twice as high.[5] Moreover, the death rate for lifelong urban residents was almost twice as great as for those who resided in urban areas less than a year.

(3) Investigators working in Nashville and Buffalo found that the stomach cancer mortality rate rose dramatically for residents of high pollution areas.[6,7]

(4) Similar findings have been reported for heart disease. In Nashville, the morbidity rates were twice as high in polluted versus low pollution areas. The mortality rates were found to be ten to twenty percent higher for residents of high pollution areas.[8]

(5) The death rates of children less than one month old is related to sulfate concentration and dustfall. Indeed, correlational studies suggest that air pollutants may be the main factor in infant mortality.[9]

With few exceptions, study after study lead to the same conclusion: Pollution causes illness. Pollution kills.

A study by Thomas Hodgson, a statistical analysis of New York City health records and air pollution levels, revealed that heart and respiratory deaths increase with rises in air pollution levels in particulate matter. Eighteen percent of all deaths due to heart and respiratory failure have been attributed to rises in the level of particulate matter. This amounts to some 30 deaths per day in New York City.[10]

Apart from its human devastation, the incredibly high cost of air pollution is much greater in loss of income to the nation than our actual budget outlays of 30 million dollars for fusion research – by at least 200 to 1! Consider the following item from *Chemical and Engineering News*, December, 1971:

> The current cost of air pollution-related sickness and premature death is about six billion dollars per year, according to the Environmental Protection Agency's Dr. Stanley M. Greenfield. This figure includes only medical care and loss of work. Damage to crops and materials is a further five billion dollars and the depression of property values in polluted areas is another five billion. Dr. Greenfield estimates that it will cost fifteen billion dollars over the next five years to halt pollution, indicating a benefit-to-cost ratio of at least five to one.

A dramatic revelation of the tragedy of air pollution is presented in the concept of "life shortening." Statistical analysis reveals that if air pollution were to be reduced by fifty percent in all major cities, then a newborn baby could expect an additional three to five years of life, lung diseases would drop twenty-five percent, and heart disease would drop ten to twenty-five percent.[11]

Whether for dollars or health, or both, the electric car has been heralded as the answer to air pollution. Since transportation produces, by weight, better than fifty percent of air pollutants, it is reasoned that a shift to electrically operated vehicles will reduce this source of pollution to near zero. This is true. The electrification of ground transportation must be regarded as a high priority item for the reduction of pollution – but the efficiency of transportation is another subject. We will cover this in a later chapter.

But if a successful electric vehicle is perfected, from what source will it derive its power? Most probably the electric vehicle will derive its power from electric batteries that will hold a charge for a long time and will be able to undergo many cycles of discharge and recharge. But how will the utilities produce the electricity for such recharging? By burning fossil fuels!

If we opt to electrify ground transportation by burning the fossil fuels to generate the electricity needed, we run into two major problems: power plants are only about one-third efficient, hence our fuel expenditure would be greater than that of direct-burning of fuel in an internal combustion engine; and this increase of fossil-fuel combustion would lead to an increase in air pollution. And, naturally, by increasing the rate of fossil-fuel combustion we decrease its lifespan, which is already perilously short. Any scheme for the electrification of transportation must consider the entire integrated system – every last component – for the production and deployment of the required electrical energy.

ARE WE CHANGING OUR CLIMATE?

A problem that is less immediately damaging than air pollution but that is more profound for the future is the problem of climate changes wrought by man.

By the year 3000 we may expect a climate reminiscent of the carboniferous epoch of 300 million years ago. At that time, the great fossil-fuel deposits were forming rapidly. The polar caps were smaller, and salt water covered what is today the Great Plains foodbelt of the United States and most of the coastal cities of the world. Will this scene return? Will it be because of the combustion of our fossil fuels over the next 100 years?

How could man so drastically, so adversely, change his environment? One scientific theory, the "Greenhouse Theory," holds that the combustion of fossil fuels will increase the carbon dioxide content of the atmosphere.[12] And since carbon dioxide is an efficient absorber of heat radiation, as the carbon dioxide concentration rises, the heat radiation emitted by the earth – in its radiation balance with the sun – will be increasingly absorbed, generating higher atmospheric temperature.

The progression will be as follows: An increase in carbon dioxide concentration will increase plant growth, not appreciably;

it will dissolve in the oceans, appreciably; but about one-half of the carbon dioxide added to the atmosphere will remain there.[13] Estimates of the ultimate effects of carbon dioxide vary, but by the year 2000 the carbon dioxide concentration may be up 50 percent, yielding an average temperature increase of a few degrees. If 50 percent of the fossil fuels are burned, then the temperature rise could be about 10 degrees, and it's a safe bet that man will burn at least 50 percent of the fossil fuels before he develops alternative power sources.[14] The temperature of the earth will rise as the concentration of carbon dioxide rises in the atmosphere, but the polar caps will melt slowly, coming to equilibrium in a 1,000-year time lag. If we burn 50 percent of the fossil fuels by 2100, then by 3100 the seas will have risen to cover the Great Plains foodbelt and all the major coastal cities of the world.

However, there is another theory. This theory holds that if the earth's cloud cover (now at 30 percent) were to increase, more sunlight would be reflected back to space and consequently the earth would become colder. The cloud cover is generated by water from evaporation of the seas and particulate matter. The particulates help in the condensation of water vapor to form a cloud droplet. These particulates are invisible, except at sundown, when they contribute to the redness of the sunset.

Which, then, will it be – cloudier and colder, or hotter? Worldwide temperature averages have actually declined over the last 50 years.

Satellite reconnaissance over the next decades may disprove or substantiate these theories. We must watch the reflectivity of the earth, its cloud cover, and its temperature variations and try to correlate these data to see if the earth is getting hotter or cloudier and cooler. Whether air pollution or some other source of dust is working subtle changes on us, it is clear, most agree, that if all variables held constant and only carbon dioxide concentration varied, the earth would get warmer. For a few degrees difference, profound changes would ensue.

Unless we discontinue burning fossil fuels, the calculations imply, we are bequeathing to a future generation a world somewhat as it was during the torrid coal-forming days 300 million years ago.

AN EVOLUTIONARY PERSPECTIVE

Whatever the rising temperature and the rising seas (or falling temperatures and seas) will demand from life on earth; or a deepening industrial sophistication, and dependency, will demand from intelligence; or what turn evolution will not take, reducing the 4.5 billion years of earth's history to a time span of a single year may help to dramatize the relative time intervals that led to our present environment and the jolt we are giving it.

If the Creation were on January 1, then by February a crust had solidified on earth. There were no oceans, only an atmosphere of methane, nitrogen, ammonia, and water vapor, but no carbon dioxide or oxygen in appreciable amounts. This primeval gas gave birth to amino acids: by the energy of lightning or solar ultraviolet light, amino acids quickly formed and then, while in the gas phase, combined to form proteins – just as they do today in the laboratory under similar conditions.

For the next two months the oceans formed by out-gasing absorbed water from the earth's mantle; a loss of only 0.03 percent of the mantle's total weight is calculated to be sufficient to create the oceans. Now the proteins (combinations of amino acids), in contact with liquid water, formed cell-like structures – again, just as they do today in the laboratory.

For the next seven months (three billion years) life evolved from the proto-cell to the first marine animals. Fermentation was the life-style in the early part of that period and produced carbon dioxide in the atmosphere for the first time – the first example of life changing its environment, even if slowly. Then, toward the latter half of this period, plants evolved – blue-green algae and others – that could use the carbon dioxide in photosynthesis: the plant kingdom was born. The plants, in turn, released oxygen, placing oxygen in our atmosphere for the first time in large quantities, thereby establishing an environment in which oxygen-breathing animals could evolve – the second environmental change wrought by life.

By November (600 million years ago) the first marine life appeared in the period called the Cambrian. The horseshoe crab is a living example of that period. Even though the oceans were in existence for at least 3 billion years, no sea-floor sediment can be assigned an age by radioactive dating of more than a fraction of that timespan, perhaps due to continental drifting – which would create fresh ocean floor and subduct older ocean floor continuously.

The common chemical origin of life is indicated by the fact that for all life – man, tree, fish, or even algae and bacteria – the DNA, the genetic material, of all species possess the very same molecules that make up the genetic code, but are arranged in different sequences, of course.

By the beginning of December (300 million years ago) the great coal and oil deposits were forming. The temperature of the earth was higher, the continents not in their present locations, and plant growth was rapid in an atmosphere five times richer in carbon dioxide than today. Some living examples of those times are the pine tree, the shark, and the cockroach.

Within thirteen days of midnight the first mammals appeared. With only twenty-three hours to go, the large carnivores of the tertiary period appeared, the woolly mammoth, for example. Then, at about two hours to midnight, or one to two million years ago, man appears. At twenty-five seconds to midnight he invents writing – about six thousand years ago.

Within the next 100 years man is committed to burn a large percentage of the fossil fuels. This is just an instant in time compared to the previous evolutionary changes. We shall be re-creating the high carbon dioxide levels in the atmosphere that existed 300 million years ago, but we shall be doing it in 100 years or so. Evolutionary forces will have no time at all to breed new species, or to breed changes into present species. The great balance among life forms on earth is presently receiving a sudden stress to which it may not be able to adapt.

NOTES

1. *Chemical and Engineering News* (December 6, 1971), p. 31.
2. "Pollution," *Chemical and Engineering News* (June 8, 1970), p. 39.
3. Lester B. Love and Eugene P. Siskin, "Air Pollution and Human Health," *Science* (August 21, 1970), p. 169.
4. P. Stocks and J. M. Campbell, *British Medical Journal,* vol. 2, 1955, p. 923.
5. C. Daly, *British Journal of Preventative Social Medicine*, vol. 13, 1959, p. 914.
6. R. M. Hagstrom, H. A. Sprague, and E. Landau, *Architectural and Environmental Health*, vol. 15, 1967, p. 450.
7. W. Winkelstein, S. Kantor, E. W. Davis, C. S. Maneri, and W. E. Mosher, *Architectural and Environmental Health,* vol. 14, 1967, p. 162.
8. R. A. Prindle and E. Landau, *Public Health Report,* vol. 77, 1962, p. 901.
9. H. A. Sprague and K. Mapton, *Architectural and Environmental Health,* vol. 18, 1969, p. 503.
10. *Environmental Science and Technology* (July, 1970).
11. Glenn T. Seaborg and William R. Corliss, *Man and Atom* (New York: E. P. Dutton and Co., 1971), p. 84.
12. Theodore L. Brown, *Energy and the Environment* (Columbus: Charles E. Merrill, 1971).
13. Gilbert N. Plass, "Carbon Dioxide and Climate," *Scientific American* (July, 1959).
14. Ibid.

Some Misconceptions About Power

"The problem is not shortage of data, but rather our inability to perceive the consequences of the information we already possess."
– Jay W. Forrester[1]

To present the facts of our impending power shortage is to evoke the most profound disbelief that such a tragedy could ever develop. Most people, out of a faith in science and an unconcern with power systems, are confirmed in their conviction that everything will work out well, that the problems are all being attended to by the right men in the right positions. They question how anyone not of this "expert" group could arrive at such private, devastating conclusions.

The public's misplaced faith in the ability of science to make a silk purse out of a sow's ear is no better demonstrated than by the number of con artists who have successfully hoodwinked financial backers into supporting harebrained schemes for making gasoline out of water by the addition of a secret compound. Indeed, it is not uncommon to hear people, even in this sophisticated age, proclaim that a cheap substitute for gasoline was discovered many years ago but that the oil companies have blocked every effort to market the product because it would put them out of business. This is sheer nonsense. But our species is particularly susceptible to this brand of chicanery. Psychologists have amply demonstrated that we often hear what we want to hear, see what we want to see, and believe what we want to believe.

Louis Enricht, a white-maned septuagenarian, understood this quite well. During World War I, he startled the world by announcing a gasoline substitute that would cost a penny a gallon! Before languishing most of his waning years in a jail cell, Enricht had duped what amounted to a "Who's Who" in finance and government, including victims Henry Ford, the editor of the Chicago *Herald*, and the directors of the Maxim Munitions Corporation of New Jersey. The efforts of foreign governments and secret agents to steal Enricht's secrets is the stuff of contemporary spy novels. Even the United States government became involved in Enricht's machinations, and attempted to obtain a court order authorizing the seizure of his factory and "scientific" papers on the basis of persistent rumors that he had sold his formula to Germany. All the while Enricht was bilking gluttonous dupes of huge sums, scientists proclaimed him a fraud. They warned the public that his claims were simply contrary to the laws of chemistry. However, their voices were unheard and unheeded.

The truth is that science can develop no energy source that nature does not possess. And no book on the impending power crisis would be complete without addressing itself to the fact that no miracles in power procurement are just around the chronological corner.

RECYCLING ENERGY

There is no such thing as the recycling of energy. Energy does pass through various modes of expression unchanged in total, but the energy passes spontaneously from a "high" (intensive) to a "low" (less intensive) form.

A ball rolls downhill. The ball is not changed, but its potential energy level is changed. When coal is burned, energy is released and work is performed as it passes through any of the following changes: chemical potential energy in the coal-air system; heat energy in the boiler of a power plant; mechanical energy in the turbine of a power plant; electrical energy in a power plant's grid; work of a machine run by electricity in the community; heat energy due to the various frictional losses. Eventually all the energy released by the burning process becomes heat. There is no net energy loss or gain, but the energy has been "dissipated" from a high potential level to a low potential level, where it can perform no more work.

HYDROGEN – A NEW FUEL?

We frequently hear talk of "inventing" new fuels. A new fuel – yes; a bigger energy bank – no! Fuels are either chemical or nuclear in nature. Either the combination of atoms releases energy (chemical) or the fission of a nucleus or the fusion of two nuclei release energy (nuclear).

What would be a new chemical fuel? Any "new" fuel derived from the fossil fuels would possess less total energy than the materials out of which they were synthesized. Expecting the invention of a new chemical fuel that will increase our total chemical energy bank can only be based upon a misconception of natural laws.

A respectable trade magazine ran a piece entitled: "Hydrogen Likely Fuel of the Future."[2] But the article put weak stress on a very critical fact: the phasing in of hydrogen as a fuel can occur no faster than the phasing in of the breeder reactor, or the development of controlled fusion, or the development of solar power. The reason is that hydrogen, unlike oil, coal, gas, or uranium, is not a primary source of energy. It requires the same energy to obtain it that it subsequently yields. To obtain hydrogen from its source, water, requires the same energy input that is released when hydrogen burns to form water. However, because all energy conversion devices are less than 100 percent efficient, there is actually a net loss.

What is so enticing about hydrogen as a fuel? Hydrogen will burn in air to produce almost zero air pollutants, since it burns to form water, a very acceptable substance to our environment. It is so clean-burning that some engineers assert that it is the only fuel that could be used in automobiles to meet the very stringent 1975-76 federal emission requirements. In addition to use in internal combustion engines, it can be employed in a fuel cell. The fuel cell would combine hydrogen with oxygen (which would yield ordinary water) and yield electricity as its power output. We'll cover fuel cells later in this chapter.

Hydrogen has many uses. We've seen that it can be used in an internal combustion engine for a car, but it can also be used as a fuel cell for an electrically driven car, or a fuel cell to provide electricity for a house, or as the energy source for a power plant, and so on. All these applications produce no air pollution. Also, hydrogen can be used to obtain many metals from their ores, expecially iron, again with negligible air pollution. Hydrogen can even be used for space heating as a substitute for natural gas.

The flexibility and benefits of hydrogen sound perfect. What's the catch? The catch is that *free* hydrogen, the gas, occurs only in trace quantities in natures. To obtain the large quantities necessary to fulfill the many potential applications mentioned above, we would have to extract hydrogen from its principal source: water. To do this requires energy. Where would the energy come from? From atomic power plants, which would be designed to electrolyze water into its constituent atoms (hydrogen and oxygen). However, our present atomic power plants are "burner-type" reactors. Their present power output is small, and, as we have seen, the National Academy of Sciences has warned that these types are consuming our precious uranium(235) isotope. They should be replaced by the breeder reactor. But we've seen that the breeder reactor will not be able to supply us with half our electrical needs until about 2010 – at the theoretical earliest – hence, it could supply us with only small quantities of hydrogen by that time. In addition, the energy bank of U(235) is far too small (about equal to oil) for the production of hydrogen from water.

And what of the effect on the lifespan of uranium should it be used to create hydrogen as a fuel? Our 660,000 tons of uranium would be reduced that much sooner.

Hydrogen as a fuel is perfect but it can't possibly be used in time to close our energy gap of the 1970s and 1980s. Massive quantities of hydrogen could be obtained from water through fusion power – but the fusion power plant has not yet been developed, and its development does not seem imminent.

NEW NUCLEAR FUELS?

How about a new *nuclear* fuel? A "new fuel" – no; a newly achieved harnessing of an "old fuel" – yes. Atomic nuclei can be used as fuels only if they provide, not consume, energy. High-atomic-numbered elements such as uranium and plutonium can fission to yield energy. In these cases, a method of obtaining the fission energy has already been developed. To create a new element in the laboratory would require very much more energy than one could ever obtain by the subsequent fission of this hypothetical new element. Such a future discovery is out of the question.

The low-atomic-numbered elements (hydrogen) can fuse to form higher-atomic-numbered elements (helium) and, thereby, release energy–as is the case in stars and our sun. But many elements have nuclei that cannot be made to either fission or fuse, with an attendant output of energy, and not for want of scientific knowledge; it is due to their nature.

The only possibilities for fusion of nuclei are for the low-atomic-numbered elements. Deuterium fusing with tritium or deuterium fusing with deuterium are examples, that we covered in the chapter on nuclear fusion. And we have also discussed fusing hydrogen nuclei themselves, as occurs in our sun. Hydrogen is not a "new atomic fuel," but, as we have seen, an existing possible fuel that has not yet been harnessed. Because of the high temperature and long confinement time required of a hydrogen plasma this prospect appears to be precluded for all "devices" except stars.

THE FUEL CELL AND THE LASER

The fuel cell has been heralded as a "breakthrough" in the generation of electricity. It has been used as a means of generating electricity on Apollo flights, to convert gas directly into electricity. In addition to its being nonpolluting, its efficiency is much higher than conventional steam-generating electricity power plants (40 to 60 percent vs. 30 to 40 percent), and will therefore cost less to operate. Because fuel cells have no moving parts, maintenance problems are kept to a minimum. Brooklyn Union Gas Company has already placed a prototype model in use. It is supplying electricity for three buildings.[3]

Is the fuel cell the answer to our future electrical needs? If gas were in unlimited supply, the fuel cell might well be a major step forward. However, widespread introduction of the fuel cell would further increase our needs for natural gas. But, as we have already seen, we are running out of natural gas and plans for synthetic gas are still in the earliest planning stages. Importing natural gas lays us open to the many dangers we have discussed throughout this book. Even if we elected to greatly amplify our foreign imports of gas, with all the associated risks, the lifespan of this source would be measured in decades. The balance of trade, in relation to oil and gas, would further exacerbate our deteriorating balance of payments position. In the light of these factors, it does not appear that the fuel cell offers real hope of solving the energy crisis, or of even making substantial inroads against it.

Often mentioned with the fuel cell as a new device for power is the laser. But the laser consumes power; it does not create it. The power output of a laser is a tiny fraction of all the power put into it. For example, if 100,000 joules of radiant energy are used to "pump up" a laser, possibly 1,000 joules of laser light emerge, emitted in one-billionth of a second. The laser's value is in the fact that its beam is the output of a trillion watts, but for only a billionth of a second!

If a laser could be built that would emit 100,000 joules in a billionth of a second and focus this light to a "pinpoint," the temperature created by the focused light on a substance would be about 50 million degrees Fahrenheit, or enough to cause fusion in a deuterium tritium fuel pellet. In fact, the laser could serve as a trigger for a hydrogen bomb. At present, all hydrogen bombs are triggered by a small uranium (or plutonium) bomb. Hence, countries with a modest technology could create hydrogen bombs, or fusion power plants, if such lasers could be built. Laser technology is presently capable of producing one-tenth the needed power density output for tritium-deuterium fusion.[4]

POWER FROM SPACE

Will we obtain power as the result of space exploration? Will solar stations be placed on the moon? Solar stations on the moon would be subjected to micrometeorite bombardment, just as they would be if placed in the planned "stationary orbit" of twenty-three-thousand-mile radius. This phenomenon of space casts doubt on the longevity of solar space stations positioned anywhere. And there is another problem: at the current payload of an Atlas rocket, it would take the equivalent of at least 100,000 Atlas rockets to build solar stations on the moon that would be large enough to meet our electrical needs.

The real value of space will come in the area of special technologies and scientific research. Because the moon has no atmosphere, microcircuit components for the electronics and computer industries can be fabricated under perfectly clean conditions which are difficult to reproduce on earth. The fabrication of special electronic components, the relaying of T.V. and other communication media by satellite, measurements of the earth's resources by infrared photography, surveillance of weather by satellite – these are space's possibilities, not power procurement.

CONSERVATION OF ENERGY

What about conserving energy by more efficient devices? Atomic power plants and fossil fuel plants operate at about one-third efficiency. It is due to the temperature range of heat input to heat output divided by the temperature of input. This means that efficiencies run from ten percent for burning wood in a steam engine, to twenty percent for coal in a steam engine, to thirty percent for high-temperature combustion, to about forty percent for very high temperature fuels. The higher the operating temperature, the greater the efficiency.

What is the maximum efficiency? The highest efficiencies might be obtained if controlled fusion is achieved – about sixty percent (ninety percent for some theoretical designs).

It will be very difficult to increase the efficiencies of the devices indicated in the following charts. By increasing the efficiency of power-producing devices (or power-consuming devices) we can expect our fuels to gain only a small percent in lifespan. Greater efficiency in power production and expression is not the answer to our impending power shortage, although every saving helps and must be implemented where possible.

CONVERSION EFFICIENCIES[5]

DEVICE	CONVERSION	EFFICIENCY (Percent)*
Electric generator	mechanical to electrical	98
Fuel cell	chemical to electrical	60
Steam power	chemical to thermal to mechanical to electrical	40
Automobile engine	chemical to thermal to mechanical	20
Solar cell	radiant to electrical	10
Light bulb	electrical to radiant	5

*These efficiencies can be raised, but only to a slight degree.

Acknowledging the problems of fossil fuel pollution from surface transportation, the efficiencies of mass transportation go far to alleviate our grinding energy shortage, especially the railroads, our most efficient carrier of weight per "energy cost." They can carry a ton of weight per mile at a far lower energy input than any other mode of transportation. Trains achieve about 200 passenger miles per gallon of fuel. The Boeing 707 and 747 obtain little more than 20 passenger miles per gallon of fuel, making the energy advantage of trains about the order of 10 to 1. Propulsion efficiencies in ton-miles per gallon fuel are airlines, 20; truck, 58; and freight train, 200. Increased speeds sharply reduce propulsion efficiencies, resulting in extravagant and unsustainable levels of energy consumption. They also add unconscionably to pollution levels. Based upon 1956 figures, the three slowest carriers (pipelines, inland waterways, and railroad freight) moved approximately *forty percent* of overland passenger and ton-miles but used up only *seven percent* of the transportation energy budget.[6] Transportation Secretary John Volpe is aware of this. He proposed making Highway Trust Fund money available for mass transit, and of the $5 billion now provided per year for highways about $2.2 billion would be diverted to mass transit systems in 1976, according to Mr. Volpe's recommendations to Congress.[7]

Two factors then will work a renaissance for the railroads: the dependency on mass transportation within our energy-starved cities, which requires an efficient carrier of weight per mile, and dependency on the railroads to ship the tenfold coal output from the mines, which we will need for synthesizing fuel.

Our present energy budget for transportation is approximately 25 percent of our total annual energy budget. Of this 25 percent, about three-fourths goes into auto and truck use.[8]

It follows, then, that if all surface transportation were conducted over rail, our energy saving would be nearly 20 percent.[9] That would indeed almost close the energy gap projected for the 1980s.

A reversal of our transportation reliance from auto-truck to rail could almost close the energy gap in itself!

However, to rebuild the railroads and extend them to the point where this reversal of transportation reliance could be effected will take decades of crash-program building; it can't be effected overnight.

And it must be accompanied by a drastic reduction in the use of airplanes, automobiles, and trucks. Naturally such readjustments will create awesome economic dislocations. Economists will have to prepare the country for a future when transportation no longer constitutes almost one-fourth of our Gross National Product (1965 figures). It is a future in which the giants of industry – automobile manufacturers and aircraft factories – will contitute a markedly reduced portion of the Gross National Product. Aircraft, incidentally, will probably be the last place where fossil fuels will be used in the future. Atomic-powered aircraft will be too heavy to fly, mostly for reasons of shielding (necessary for protection from the power plant), and also because of the great weight of even the smallest power plants. Perhaps the future might see giant atomic-powered aircraft, but small planes seem to be precluded from nuclear power on fundamental principles.

GARBAGE DISPOSAL AND MATERIALS RECYCLING

How much power can we get from garbage pyrolysis? Pyrolysis (*pyro* = heat or fire; *lysis* = a loosening or releasing) is a process in which organic compounds are heated in an airless chamber. In the case of garbage, this yields three major products – a combustible solid, a combustible oil, and a combustible gas (plus a solid residue). Pyrolysis of wood, for example, yields wood alcohol and other chemicals plus a residue of carbon that is used as a filter in gas masks and as a general bleaching agent.

The theoretical power yield from garbage is small. It amounts to a potential of about three percent of United States' 1970 electrical needs. However the great advantage in pyrolysing garbage is that the energy obtained from the products of pyrolysis more than compensates for the energy needed for this disposal method. The disposal of garbage by pyrolysis is a developing practice, especially in Europe, due to the net gain of power from the entire closed-loop process.

On a national basis, the potential power gain from combusting the pyrolysis products of garbage are approximately:[10]

(1) United States tonnage of garbage = 160 million tons per year.

(2) At the usual efficiency for steam electric generation, its potential wattage equals 10 billion watts.

(3) 1970 United States electrical output = 375 billion watts.

The problem with power from garbage disposal (or animal waste) is that the material is dispersed over the entire nation. Hence, collecting it would take a high energy cost. And in garbage's low efficiencies in conversion into a fuel and then into power, it reduces a modest, but inadequate, energy source to a very inadequate energy source.

Will the recycling of materials save energy?

Perhaps a little saving can be effected by recycling; perhaps not. The extent of recycling of some basic materials in our economy is as follows: lead, 50 percent; steel, 45 percent; copper, 40 percent; zinc and aluminum, 35 percent; and paper, 20 percent.[11]

A rule of thumb for the steel industry is that it takes a ton of coal to keep a ton of steel in use – whether one starts with iron ore or scrap steel. Of course each industry faces different recycling technologies, but the largest part of our energy output goes to uses other than manufacture, as follows: industrial, 37.2 percent; transportation, 24.6 percent; residential and commercial, 22.4 percent; heat loss, 15.8 percent.[12] The real advantage in recycling is to conserve our nonrenewable material resources, not to conserve energy.

OUR ESTIMATES COULD BE WRONG

Perhaps we will happily learn one day that our energy resources have been underestimated by the experts. But that works both ways.

The key to estimating the total extent of a resource is to exploit one small area to find out how much oil or how many tons of copper per volume of earth it yielded, and then extrapolate to all similar sites on earth. We cannot extrapolate to the total area of the earth, for not all areas are "replays" of other areas.

For example, not all areas on earth have experienced the same history of soil bacteria; hence not all areas are comparable in resource content. Many resources came into being as a result of a life form that lived millions of years ago. Iron ore deposits are believed to have been caused by bacteria that employed iron in their biological energy supply systems. Sulfur deposits came from anerobic ("no air") bacteria that used hydrogen sulfide as "food" for energy production. Coal and oil are the partial decay products – under anerobic conditions – of plants. Soil bacteria have, over the millennia of our earth's history, flourished and perished, species have come and gone, and in the process deposits of phosphates, iron, sulfur, fossil fuels, and many other resources were left behind. Soil bacteria play a profound role in nature, even today, for they fix nitrogen from the air and thereby provide an indispensable nutrient without which the entire plant kingdom could not exist, nor the animal kingdom which is parasitic upon plants. Over the earth's history, the land exposed to the air and then inundated by sea water has always changed, and with these changes have come a reshaping of microscopic life forms, and hence a reshaping of our resources.

In addition to soil bacteria as causes of resource formation there are inanimate causes: the circulation of ground water, the differential solubility of salts as temperatures fluctuate, and so forth.

The estimates of resources given in this book could prove to be low, as well as high. For example, the world's reserves of oil contain a 1,200-billion-barrel component called "presumed." Maybe some of it won't be found. Also, we have seen that the most available part of a resource is taken first, then the less accessible parts are sought, so that an estimation of a resource's total amount is devoid of an indication of how hard it might be to get it all.

However, the margin of resources left to us is so small that reliance on the testimony of experts being wrong is carrying optimism to be verge of indifference. We cannot afford the luxury of testing the predicted limits of our resources only to find that the shortages – and the crisis – have overwhelmed us.

NOTES

1. From *Technology Review* (January, 1971).
2. *Chemical and Engineering News* (June 26, 1972), p. 14.
3. "Fuel Cells Light 3 Buildings Here." *The New York Times,* June 28, 1972.
4. "Fusion Scientists Are Optimistic," *Chemical and Engineering News* (December 20, 1971), p. 34.
5. *Energy and Power, Scientific American* (September, 1971), p. 151.
6. Richard A. Rice, "System Energy and Future Transportation," *Technology Review* (January, 1972), pp. 35-36.
7. "Away from Highways," *Time* (March 27, 1972), p. 90.
8. Calculated from Richard H. Rice, "System Energy Is a Factor in Considering Future Transportation," ASME meeting, December 1970. In the early 1960s, automobiles and intercity trucks used 33 billion gallons of fuel out of 42.5 billion gallons employed in overland transport. This comes to 77.6 percent.
9. Assuming 42.5 billion gallons employed in transportation and that all could be carried with the efficiency of freight trains averaging at least five times greater than auto and truck combined, we would save some 34 billion gallons per year, or 80 percent. When multiplied by 25 percent (representing total energy employed in transportation) the savings would be 20 percent of the total energy budget.
10. U.S. Bureau of Mines.
11. "Solid Waste," *Environmental Science and Technology* 5 (July, 1971), p. 594.
12. *Energy and Power, Scientific American* (September, 1971), p. 224.

A Comparison of Nations

We have referred to the growth potential of the Russian economy and the threat of no growth – or collapse – for our own economy. Which of us is most likely to win world leadership?

What is the likely future power posture of South America, Africa, China, Western Europe, Japan, the Middle East, Canada, and Russia?

Which nations can expect to rise to predominance in living standards and in military might? And how long will their power base serve them? That is, which nations can rise to "superstatehood?"

Not only are resources important (energy, mineral, and human), but so is the timing of power-base transformations.

RUSSIA: THE ONLY SURVIVING SUPERSTATE

Russia not only has a vast natural resource base in timber, water, iron, coal, and minerals, but she has these in high proportion to her 240 million people. Of all nations, Russia is by far the most self-sufficient in all the natural resources that go into a modern technology.[1] Also, her research into power technology is foremost in the world. As we have seen, Russia has made great breakthroughs in magnetohydrodynamics (MHD) and in fusion research.

Just how well off is Russia in terms of energy resources?

TOTAL RESOURCES[2]	RUSSIA	UNITED STATES
Oil (billions of barrels)	400	100
Gas (trillions of cubic feet)	3,000	1,000
Coal (billions of tons)	4,000	1,500
Hydroelectric (billions of watts)	240	160
Uranium (tons)	1,000,000	660,000

CHINA, RUSSIA, AND THE
UNITED STATES–POWER PRODUCTION[3]

DOMESTIC PRODUCTION OF	U.S.	U.S.S.R.	CHINA
Coal (millions of tons)	600	690	300
Oil (billions of barrels)	4.2	3.2	0.15
Gas (trillions of cubic feet)	23	7	negligible
Electricity (billions of watts)	375	170	16
Population (millions)	210	240	800

(1) Russia faces an oil shortage in thirty to fifty years – not "now" as we do.[4] Russian oil consumption is about sixty percent of ours, at present, but Russia has about four times our oil reserves. Her proved oil is less than ours, but statistical considerations indicate a large presumed component of her total reserves. At the present rates of domestic oil production, Russia will surpass the United States by about 1978, at which time Russia will be producing about five billion barrels per year of domestic oil.[5] While our capacity to produce oil is leveling off, Russia's is rising – at about eight percent per year.[6] After Russia surpasses us in domestic oil production by 1978 she will still have about four times our oil reserves yet remaining.

(2) Russia has about three times as much natural gas as we do: 3,000 to our 1,000 trillion cubic feet. Most of the Russian gas is in central and Asiatic Russia, in the Tyumen province, not in her industrial west (east of the Ural mountains) where it is needed. Russia's need to ship gas over 1,000 miles to European Russia has led to her research into large-diameter pipeline transmission systems that transmit gas under high pressure. Our largest proposed pipeline will be about 42 inches in diameter; the Russians use 40-, 48-, and 56-inch pipe, and are experimenting with 99-inch pipe. They have just signed a multi-billion-dollar deal with Occidental Petroleum to provide technical help in the development of their oil and gas fields.

Russia's gas consumption rate is about 7 trillion cubic feet per year. Compared to her gas reserves she has over 400 years of gas! Of course, Russia is consuming gas at an ever growing rate, just as we are, but her gas shortage will only come well after the year 2000 – ours has already arrived.

(3) Russia has three times our coal reserves: 4,000 billion to our 1,500 billion tons. Again, most of Russia's coal is in central and Asiatic Russia about, 1,000 to 2,000 miles from where it is needed. The fraction of coal that is burned to generate electricity is mostly burned by power plants located near the coal fields. The long distances between the power plants and the consumers has led to a pioneering effort to develop efficient long-distance transmission

lines. For example, a direct-current 1,500 kilovolt line will connect the power plants in the Ekibastuz coal fields with the cities of the west.[7] Transmission of direct current will probably be the chosen technique should fusion power ever be developed; hence, this technology has a place in the future. In contrast, we use alternating current transmission lines, which are much less efficient.

By their interest in coal we can see how the Russians are reacting to the breeder reactor's being many decades off. She has contracted with Joy Manufacturing, an American company, for automatic mining equipment.

(4) The Russians won't have to develop synthetic fuels from coal because her oil and gas reserves will provide her with the time to research fusion power. Should Russia's oil be in short supply by the 1990s, she can use natural gas as a substitute.

(5) Russia's hydroelectric potential is about 1.5 times ours. Russia's electrical output has risen from 16 billion watts in 1951 to her present 170 billion watts; ours is 375 billion. Russia's electrical output is about 70 percent of her hydroelectric potential, which is 240 billion watts; ours is more than twice that amount.

The Russians have forged ahead in many areas: 50 percent efficiencies in the generation of electricity from coal (magnetohydrodynamics; efficient electrical transmission lines (direct-current transmission); and high-pressure and large-diameter gas pipelines. Where they appear to be in no great hurry is the technology that cannot be crash-programmed due to the laws of physics – the breeder reactor, with its inherent fuel-inventory breeding requirement. Ironically, our present research budget for power development places the greatest future reliance on power from the breeder reactor.

Let's compare Russia's fusion efforts with ours.

(1) The United States has spent a cumulative sum of $400 million on fusion research to date; the Russians have spent over $800 million!

(2) Fifteen percent of all fusion scientists in the world are Americans; 35 percent are Russians.[8]

(3) The most advanced fusion reactor in the world is Tokamak:[9] a toroidal configuration developed by the Russians at the I.V. Kurchatov Institute of Atomic Energy in Moscow in 1969 and subsequently followed by the United States in similar reactors at Princeton, M.I.T., and Oak Ridge National Laboratory.

The Russians could achieve a higher living standard than ours in just two to three decades. Based on our fuel self-sufficiency being down about 30 percent by the 1980s, a nuclear capacity not ready until after 2010, synthetic fuels not ready until after 2000, and our having to work hard to just stand still, that is, to just partly fill the energy gap, if we manage to avoid an industrial collapse altogether, our economy will level off – at best – perhaps even registering a gain of about 12,000 to 15,000 watts per person "living standard."

A slightly higher living standard of 12,000 to 15,000 watts per person by 1980 is not exactly equal to our nation's goal of uplifting the disadvantaged, and indeed everyone, but it's not the tragedy of a 30 percent decline in power. The point is that we face a depression of 1929 scope, or at best holding our own, as future prospects, whereas the Russians face the happy prospect of a rapid growth in energy deployment each year with no shortages developing until well after the year 2000.

The Russian energy deployment rate is approximately doubling every ten years: from 1960 to 1970 their rate of oil, gas, and coal consumption has doubled, and their rate of hydroelectric development has doubled. Therefore, their energy consumption growth is about seven percent per year – and they have the resources to sustain it. Whereas this growth rate approximates ours, we're "topping" out.

Hence, by the late 1980s or thereafter Russia may be a military superstate and the proud possessor of the world's highest living standard. We may have lost our image and our capacity to influence world events; in short, lost the cold war.

Absurd? Just work out a Russian seven percent rate of growth, compounded annually, to a zero growth – or decline! – for the United States (after 1975) for about twenty years, and this reversal of status among nations becomes evident.

Based on present trends, by 1985 Russia could have a living standard of 15,800 watts per person from domestic sources; the United States can expect only 11,000 from domestic sources.

No one can fully foresee the ideological consequences of the developing energy crisis. But it's hard to believe that profound changes in power-politics will not follow the profound national changes in industrial power now working their way to the surface of events.

CHINA: AN ENERGY PYGMY

China, which is striving toward a military superstate, will never be an economic superstate because of her low energy resources, especially as seen on a per capita basis. With four to five times our population, China has two-thirds our coal reserves and one-tenth (discovered) our oil reserves.[10] She imports oil from Rumania. Even though China possesses a rapidly developing technology, she lacks the energy reserves within her borders to produce a mighty economy to support a high living standard for her 800 million people.

China will undoubtedly influence world affairs, but her people cannot even aspire to present United States living standards. Her population is 80 percent agrarian and mostly unmechanized at that. If China could mechanize her agriculture the flood of people to cities would engulf her.

Manpower is one of China's major resources; farm industrialization is not possible within one generation, nor is it desirable. If she mimics the West's or Russia's industrial complex she is doomed to distant second-rate standing. But she aspires to world leadership.

Data on oil, gas, coal, water power, and uranium for China is scanty, but the figures indicate that China is weaker than the United States in fuels. But with four to five times our population, and combined fossil fuels estimated at about one-half to two-thirds that of ours, one can see their ultimate weakness. Hypothetically speaking, if China's 800 million people were to draw 10,000 watts per person (distributed as in the United States among coal, oil, and gas) from coal (since her oil and gas reserves are small), her coal reserves would last under a century. A sophisticated living standard for all China's people cannot be achieved "Western style": they will have to devise a unique quality of life in order to compete abroad with the West or Russia in national image and prestige.

China's per capita energy expenditure can never match that of Russia's or the West's.

FOSSIL FUELS RELATIVE TO POPULATION[11]

	U.S.S.R.	U.S.	CHINA
Oil (barrels per person)	1,700	500	110
Gas (1,000 cubic feet per person)	13,000	5,000	unknown
Coal (tons per person)	17,000	7,500	1,200

China, not surprisingly, claims parts of Siberia where Russia has much of her fossil fuels! This is the area of Russian-Chinese border disputes.

If China actually held the area she claims, her fossil fuel situation could change dramatically.

CANADA: RELATIVELY STRONG

Canada has one-tenth our population, but only one-twentieth our coal reserves. However, she has one-third our oil and one-fourth our hydroelectric potential.[12] Relative to her population, Canada has three times our gas and oil, and therefore she would not face an oil and gas shortage until about five to seven decades, provided she used oil and gas at the same per capita rate as we do, and did not export to us! We have already mentioned that

Canada's Energy Commission has discussed the need to restrict fuel exports to the United States. That is, Canada is more energy self-sufficient than the United States – but not to an immortal degree, by any means.

Canada, however, does not possess the high technology of the United States, Russia, or the United Kingdom. Whereas she can produce a very high living standard for her people, on military and population grounds super-statehood is unlikely for the immediate future. Russia and Canada are in the same favorable energy position on the basis of energy resources per capita. However, even without her military strength, just from an energy standpoint, Russia's vastly higher technological base enables her to dwarf Canada as a pacesetter in world affairs.

THE MIDDLE EAST: SUDDEN POWER

The political and economic clouds that hang over the Middle East grow dark, and will grow even darker until a climax is reached in the 1980s, when the sheikdoms of the Persian Gulf will hold in their hands the power of oil-attendant life and death over Western civilization. They will wield astonishing influence over Japan, Western Europe, and America, and we will have permitted it to pass.

How did we engineer caesars out of sheiks? By neglecting to develop a synthetic fuel capacity during the 1950s and 1960s! We have seen how we will become dependent upon imported oil in the mid-1980s!

The Middle East sheikdoms have no desire to see the nations of the Western world develop a synthetic fuel capacity. They have little coal, iron, aluminum, timber, or any other key resource for the construction of a modern state. They are without trained technical personnel. Without oil to sell they have nothing. So their future is also clear: they will inherit enormous power as the United States, Japan, and Western Europe bid for their oil in the mid-1980s. They can put their oil up for auction to the highest bidder. They are and will continue to demand high oil prices. Hopefully the nations of Western civilization will develop a self-sufficient synthetic fuel capacity, or fusion power so that the Middle East will be without leverage. For them, it will mean historical oblivion, so far as shaping international relations goes; the Arabs have just two decades to strike while the iron is hot.

In the meantime, Israel believes that a pool of oil exists under the northern Sinai desert, land captured during the Six-Day War. Preliminary oil finds, discovered since then, have turned Israel from an oil-importing to an oil-exporting state.

How much oil will be found under the Sinai no one knows for sure. But suppose, as some experts believe possible, that the oil fields prove to be even one-tenth of the oil fields of the Persian Gulf? Then the oil fields are indeed significant; that one-tenth could be as large as one-half of all United States domestic oil fields!

Of course, the struggle in the Middle East proceeds on themes historically independent of oil, but this new oil development could certainly add to the conflict in unpredictable ways. The amount of oil under the Sinai will influence the degree to which the Arab world holds our oil lifelines in its hands during these next two critical decades. Russia and Western Europe will be watching closely.

LATIN AMERICA

Latin America's population is about 300 million. Her ultimate resources of oil, though variously estimated, are at best about 200 billion barrels (mostly from Venezuela) – twice that of the United States. Her coal reserves are estimated at about 20 billion tons – about 1/75 of the United States.[13]

What does this energy picture suggest? Latin America consumes about 1 billion barrels of oil per year. At this rate, she could be oil-sufficient for 200 years – but at a continuing low living standard. However much she raises her living standard, so would her oil position deteriorate even more. For a United States type of economy her oil reserves would last only thirty to forty years. And her capacity for synthetic fuel production from coal is worse than Europe's. After she burns her oil she cannot turn to synthetic fuels as a substitute.

Theoretically Latin America has about thirty years in which to go nuclear, but she lacks the uranium and the trained manpower. Latin American countries, along with many independent states in Africa, will linger in the fossil-fuel epoch longer than most countries.

AFRICA

Africa has large oil reserves in Libya, Algeria, and Nigeria. Large reserves, that is, relative to Africa's present oil needs – but small when compared to the reserves of the Persian Gulf. Africa's oil is shipped primarily to Europe. Her uranium reserves are primarily in the Union of South Africa; her copper is mostly in Zambia; her hydroelectric capacity mostly in Egypt and the Congo.

What could become of Africa? Her several nations, either alone or collectively, do not possess the fossil-fuel energy banks to give her 344 million people even a European living standard. Her coal is only 100 billion tons (about 1/15 that of the United States).[14] Her uranium reserves are probably as large as those in the United States but she does not possess the technological base to realize this potential energy source. Even in the Union of South Africa, the technological base is insufficient.

Africa's copper reserves are estimated at less than those in the United States. Hence, if we face a copper shortage before the turn of the century when we will have a four-trillion-watt economy, surely African copper would be insufficient to sustain them at any comparable living standard.

Africa faces the problems of (1) insufficient oil and gas reserves for a high living standard; (2) a coal-to-population ratio worse than Europe's or similar to Latin America's; and (3) insufficient technological manpower to develop her uranium and copper resources (both in short supply) for a nuclear age.

However, Africa has a large hydroelectric power potential – about 1.7 trillion watts.[15] This means that the minimum per capita power available to the peoples of the continent is about 4,000 watts per person. Of course, the mighty rivers of Africa are located in different geopolitical regions – the Congo and Egypt. It takes time to develop this potential and an enormous electrical capability to deploy this power. But even resolving these factors, the hydroelectric potential could be manifested in only a modest living standard: 4,000 watts per person compared to 10,000 in the United States today. Whereas it is not trivial, it is remote. Also Africa is a politically divided continent. The "Arab North" is quite a separate entity from the "Black South," which is largely white-ruled. Long before any plans to realize the potential of the re-souces of the southern half, we can expect political upheavals that will change Africa's social structure.

JAPAN: THE MOST VULNERABLE OF NATIONS

Much has been written of late characterizing Japan as "the biggest economy by 2000." This is based on a dramatic growth in her Gross National Product (GNP) since 1950. What is the nature of the Japanese economy and the basis for an almost universal optimism for it?

Japan's corporate structure is privately owned. However, its security market (stock market) is weak. Corporations obtain funds for expansion through private banks, which in turn draw from the Bank of Japan. The Bank of Japan, in turn, makes its decisions based upon discussions between government officials and private industry. The government is represented both by the Ministry of International Trade and Industry (MITI) and the Ministry of Finance; industry is represented by the Zaibatsus–huge conglomerates that include distribution companies and banks as well as manufacturing companies.

In effect, the government selects which industries it will financially aid and then stands behind the loan. The government permits a high-debt-to-equity ratio to be incurred – 5 to 1 (in the United States it is 0.5 to 1). By this financial strategy, the Japanese have invested heavily in industries that are efficient and that have future growth potential. Some of their more outstanding examples of success are in the areas of microcircuitry (radios, telecommunications equipment, and the like), supertankers (Japan is the world's largest shipbuilder), automobiles (Japan ranks behind only the United States), steel (Japan is slightly behind the United States and U.S.S.R.), and computers (Japan ranks only behind the United States).

It is not true that Japan "exports to live," as many believe. Japan's production is mostly consumed at home. The percentage of her production that is exported is actually less than in the European Economic Community. The United States exports 13 percent, Japan 30 percent, West Germany 37 percent, England 48 percent, and Canada 63 percent. Japan must "import to live"; she imports the following percentages: coal 80 percent, iron ore 98 percent, and oil 99 percent.[16]

Japan has enjoyed a very favorable tariff arrangement with the United States and other countries. She has raised great taxes on goods coming from the United States while we have imposed only nominal taxes on Japanese goods. This arrangement has its roots in our economic policy following World War II; we decided to strengthen Japan by giving her easy access to our domestic markets. The result is typified by the fact that Japanese autos are imported by the United States at nominal taxes whereas a United States Pinto automobile that costs $2,000 in the Unites States costs $5,500 in Japan. The story is similar for other products. Strengthening Japan economically was held to be in our national interests for world political stability.

A unique feature of Japan's social structure is that men and women are employed on a lifetime basis by one company. The company is willing to pay for retraining, and the worker does not resent automation – he is simply retrained. Each worker has an income related to his age, and so has no fears about job reassignment within his company. The mandatory retirement age for workers is 55, but not for management, and so only the most adaptable and talented people are retained after 55 years of age. The Japanese labor force is 25 percent unionized, as in the United States, but it is unionized by company – not by trade or occupation. Therefore, the union of a company and a company's management have the same long-range goals. Of all industrialized nations, time loss due to strikes is the lowest in Japan.

The Japanese industry-government relationship is unique among industrialized nations: it is not Western capitalism nor Eastern Communism. Some call it state capitalism; others call it a technocracy; still others call the entire nation "Japan, Inc." Whatever it is called, the very agile and foresighted government-industry partnership is very productive of a rapidly rising living standard.

But consider Japan's energy situation. She has almost zero gas and oil. Her coal reserves are one-tenth England's or about one one-hundreth that of the United States, but she has 100 million population.[17] Japan has little iron or other basic elements. Without the importation of oil from the Middle East and coal from America, she would be out of energy and out of business.

Obviously Japan cannot pursue a domestic (self-sufficient) synthetic fuel strategy. Nor can she pursue a domestic nuclear capacity because there are no uranium reserves. She must purchase enriched uranium from the United States – and we need our uranium at home.

Apparently, Japan can't rely on Middle East oil forever. As we have indicated, in just a few decades, barring political acts of oil cutoff, the Middle East's oil reserves will be severely depleted. Japan must switch from Middle East oil dependency to United States dependency, either for coal to synthesize oil and gas or for uranium when and if Japan can "go nuclear." However, this will not happen soon, if at all, considering our own shortages.

What is Japan's solution? To develop controlled nuclear fusion, which could provide unlimited industrial power for all nations. Judging by their many technological accomplishments, they have the personnel to do it, but that is in the distant future.

The biggest economy by 2000? On energy grounds alone

this is preposterous. Where will Japan get its energy sources between now and 2000? From its competitors, who are also in short supply? A more reasonable guess about Japan's future is that she is in for a threatening period of energy shortages if not an energy blackout altogether. Japan in the middle 1980s is looking forward to about 8 percent of her oil needs to be supplied by Russia, a maximum of 20 percent from Indonesia, and 8,500 tons of enriched uranium from the United States. Our future capacity to produce enriched uranium for ourselves and Japan has been projected by experts to be insufficient.[18] Obviously, another preposterous game-plan by all nations concerned.

In April, 1972, then Premier Eisaku Sato confirmed Japan's plight when he told a *New York Times* reporter that for Japan, "power is the key – in oil and nuclear power – for the next thirty years."

Japan is pinning some of its hopes on the Senkaku Islands, which lie between Taiwan and southern Japan. They were previously administered by the United States but returned to Japan on May 15, 1972. Even though there was a formal Japanese takeover there still remains a lingering dispute with China and Taiwan (Formosa). Both claim the Senkaku Islands; both want the oil that is presumed to be offshore of these islands.

If the oil fields prove to be large, Japan stands to gain a temporary strategic energy advantage over China, who will continue to remain far behind her rival Russia.

WESTERN EUROPE: MORE PRESSING THAN IN AMERICA

The countries of Western Europe will probably face an energy future similar to that of the United States, with one exception: if they are denied oil from Libya or from the Persian Gulf, they face an immediate and total industrial collapse because Western Europe has only meager alternative oil reserves, in the North Sea, or about five years' worth at 1970 consumption rates. In Western Europe, the vast majority of the coal reserves are possessed by West Germany (about 62 percent) and the United Kingdom (about 36 percent). All of Western Europe's coal reserves are about equal to one-quarter those of the United States. Their oil reserves are only about 20 million barrels, and West Germany possesses 75 percent. France ranks significantly behind West Germany and the United Kingdom in fossil fuels.

What can Europe do? She will probably develop a synthetic fuel strategy and thereby live off her coal reserves. But we have seen that this strategy is good for only a few decades, that it is a monumental technological undertaking, and impossible to effect quickly. To make it safely into the nuclear age, Europe must phase out of foreign oil and into synthetic fuels within twenty years and concurrently develop a breeder-reactor capacity along with the electrical requirements of a nuclear age.

But Western Europe doesn't have the copper or the uranium for a nuclear age. Europe is in the same predicament as Japan – except that Europe has coal, Japan does not. Even if Japan and Western Europe make it into the fusion age with us, where will they get the copper to conduct electricity? Most copper reserves are in North America, Chile, and Zambia. We can expect global maneuvering for copper ore, as we've begun to see in Chile. The struggle for copper will add to the strain of peace among nations.

The countries of the European Economic Community have no uranium enrichment facilities or uranium reserves. At present, a consortium is in the planning stages to construct the vast facilities necessary to enrich uranium. After this is accomplished Europe must look for uranium to enrich. From where? The Union of South Africa appears the most likely source for uranium ore. Russia, in her desire to hold the power strings in Europe in a nuclear age, has already begun discussion to sell enriched uranium to Western European countries, which presently purchase enriched uranium from the United States.

Controlled nuclear fusion, once again, appears to be the ultimate answer for Western Europe, just as it is for Japan and utlimately for every nation. Water, with its deuterium percentage, is everywhere, and each gallon could yield the equivalent energy of 300 gallons of gasoline.[19]

AN INTERNATIONAL CENTER FOR ENERGY PROCUREMENT

The following analogy, comparing fossil-fuel energy reserves to populations, will outline the relative strengths of the United States, China, and Russia. If America is six feet tall, Russia is twelve and China is one (two, should she gain her claim in Siberia). By further analogy, Latin America equals an inch. Such "heights" define the relative fossil-fuel energy banks on a per-capita basis.

World leadership will not necessarily go to the superior culture, but it may go to the superior technology. Freedom of press, religion, or speech, or a choice of jobs or even freedom to change the system may not be the final influence on world events. Perhaps the energy to produce a strong material environment – military and domestic – may prove to be the right kind of international influence.

Russia, in her vastly superior energy position, will prove to be a very formidable foe in these next twenty years. We are passing into a very threatening epoch in American history, in which we could become a second-rate power along with an equally diminished free world.

As we have seen, all nations have a vital stake in fusion power – either they don't have sufficient fossil fuels, or sufficient uranium reserves, or both. All nations must turn to fusion power – it's just a question of when.

Those nations most immediately affected by the energy crisis, Japan, the European Economic Community, and the United States, should form a consortium to discuss its many-sided problems. Such a consortium for formulating the scientific, economic, and political strategy could research all aspects of power procurement and deployment, from synthetic fuels, to controlled nuclear fusion, to the electrical requirements of a nuclear age.

We should not wait until natural forces, or a foe, plays one against another for the scant energy sources we all draw upon. Perhaps we could form an international center for energy procurement – INTERCEP– and invite all nations, including Russia and China. There is already a precedent for the invitation of Russian scientists (who constitute 35 percent of all scientists working on fusion):[20] we cooperate on fusion research to a small degree. Our fusion reactor at Princeton has been remodeled after the Russian reactor at the Kurchatov Institute in Moscow after a Russian breakthrough in 1969 on plasma-confinement magnetics. Cooperation between the U.S.S.R. and the U.S. on fusion research proceeds without a formal treaty. It is a cooperation among scientists whose work cannot be translated into weapon systems, and thereby the data is not classified.

A more formal partnership among nations is the European Atomic Energy Community (Euratom). This is composed of members of the EEC, Belgium, France, Germany, Italy, Luxembourg, and the Netherlands. Other nations engaged in fusion research are Japan, the United Kingdom, and Sweden.

Hopefully, the energy crisis, in the hands of statesmen, may finally bring about international cooperation.

NOTES

1. Peter G. Peterson, "The United States in the Changing World Economy," vol. 2 (Washington, D.C.: Government Printing Office).
2. Adapted from: Committee on Resources and Man, National Academy of Sciences, National Research Council, *Resources and Man* (San Francisco: W. H. Freeman and Co., 1969), ch. 8 and *Energy and Power, Scientific American* (September, 1971).
3. Adapted from: Central Intelligence Agency, "People's Republic of China Atlas" (Washington, D.C.: Government Printing Office, November, 1971), no. 4115-0001.
4. "Oil-Producing Lands Weigh Nationalization," *New York Times*, June 6, 1972, sect. 3, p. 1.
5. *New York Times*, June 6, 1972, p. 66.
6. Ibid.

7. "Soviet Approves Asian Power Project," *New York Times,* July 3, 1972, p. 51.
8. "The Quest for Fusion Power," *Technology Review* (January, 1972), p. 10.
9. Bruno Coppi and Jan Rem, "The Tokamak Approach in Fusion Research," *Scientific American* (July, 1972), pp. 65-75.
10. Central Intelligence Agency, "People's Republic of China Atlas."
11. Derived from data given in this chapter.
12. *Energy, Life Science Library* (New York: Time-Life Books), p. 193.
13. *Energy and Power, Scientific American* (September, 1971), pp. 64-65.
14. Ibid.
15. Committee on Resources and Man, *Resources and Man,* ch. 8, p. 209.
16. Peterson, "The United States in the Changing World Economy."
17. *Energy, Life Science Library,* p. 193.
18. "More Enriched Uranium Capacity Needed," *Chemical and Engineering News* (August 17, 1972).
19. Glenn T. Seaborg and William R. Corliss, *Man and Atom* (New York: E. P. Dutton and Co., 1971), p. 43.
20. "Fusion Scientists Are Optimistic," *Chemical and Engineering News* (December 20, 1971), p. 37.

The Politics of Power – International

RUSSIAN STRATEGY

We have considered that Russia need only wait to achieve her supremacy. She can, if she chooses, be more active, entering into negotiations with energy-impoverished nations and thereby start creating worldwide dependency upon her for vital resources.

The intricacies of the power crisis among nations will draw us toward conflict with Japan and Western Europe. By the mid-1980s we will be fiercely competing for the same Middle East oil. Even so, buttressed with a synthetic fuel capacity in the United States, we might well have to continue to provide Japan with coal, and Western Europe thereafter, including supplies of uranium for a nuclear future. Even if we manage to develop the breeder reactor and the plutonium fuel, we will be no less hard pressed to share our scant fuels with our allies.

China and Russia will be tempted to intervene in the Middle East to worsen our position, while strengthening their own. They might even try to cultivate such oil-rich countries as Canada, Venezuela, Nigeria, and Indonesia, which would force the West to take countermeasures and to endanger world peace.

Consider the following recent developments:

(1) Russia recently concluded agreements with West Germany, France, and Italy to supply Siberian natural gas by the second half of the 1970s.[1] Is this a first step in a coordinated monopoly on

energy in Europe? Russia will export a maximum of 400 billion cubic feet of natural gas per year, whereas she has signed contracts with Afghanistan and Iran to import 479 billion cubic feet per year![2] This makes the U.S.S.R. a net importer of gas even though she has the world's largest reserves.

(2) Russia has proposed to Japan to build a 4,000-mile oil pipeline from Tyumen (a rich oil field) to Nakhodka (a seaport facing Japan). The Japanese would pay $1 billion of the $3 billion cost in return for a 20-year supply of up to 35 million tons of oil per year (17 percent of 1972 needs).[3]

(3) At recent high-level meetings between Soviet and Yugoslav economic specialists, the Yugoslav delegation is reported to have negotiated accords that will provide Belgrade with sizable supplies of petroleum and coke.[4] In addition, Russia has also entered into negotiations with energy-rich nations in what appears to be an effort to gain control over the "energy flow" of the world. To illustrate:

(4) The Soviet Union recently concluded an agreement with Libya to "jointly develop and refine Libyan oil. Tass said that the agreement 'provides for cooperation in prospecting, extracting and refining oil, in developing power generation and other branches of Libya's economy as well as prospecting for gas, and includes training Libyan national cadres.' "[5] The agreement represents a major breakthrough against the Western monopoly over Libyan oil.

We might consider what would happen if Libya signed a contract that permitted Russia to purchase all 2.4 billion dollars worth of Libya's oil per year. That's 1.8 billion barrels per year, about 42 percent of Western Europe's 4.7-billion-barrel oil budget per year (about 6 billion dollars worth).[6] If Russia could purchase just 20 or 30 percent of this total, or contract to curtail its production (similar to the soil bank program in the United States, in which farmers are paid not to produce), she can strangle Western Europe.

It is not impossible. The Arab world wants to be able to purchase technology with its oil money – obtained by the sale or the nonproduction of oil – for factories, power plants, and other modern developments.

The key to their diabolical strategy is that the Arabs would consent to such a deal only if they could be guaranteed free purchase from Russia of all the means to produce a modern technology. They don't want Communism, or capitalism, but modernization. And we will not find solace in looking at previous decades when this strategy would not work. Then, because oil needs were much less, reserves were much greater, and Western Europe could simply buy United States oil.

(5) President al-Bakr of Iraq recently announced an accord with Russia. Under the terms of the agreement, Iraq will now develop her oil potential with Russian assistance. Iraq had nationalized the Iraq Petroleum Company (a Western-owned consortium) and thereby ended further Western exploration and development of its fields.[7] The parent companies of the consortium just nationalized by Iraq control 60 percent of the world oil market and they have retaliated by boycotting Iraq's oil. This will drive Iraq further into the arms of Russia.

In a related development, Syria announced a plan to nationalize the assets of the Iraq Petroleum Company on her territory.

Although the majority of oil-producing countries in the Middle East remain pro-West, tremendous political and, possibly military pressures will be applied by Russia and the U.A.R. against these countries to divest themselves of Western interests. Nationalization of oil is a first step. Already, Libya, Iraq, and Algeria have nationalized (at least in part) the assets of Western-owned consortiums that develop their oil and gas reserves. Is this a first step to a "pro-" and "anti-" West polarization?

The following are the proved reserves of oil (in millions of barrels) from the largest oil-rich nations in Africa and the Middle East (those marked with an asterisk have nationalized all, or part, of the assets of foreign oil companies in their lands): Saudi Arabia, 85,000; Kuwait, 68,000; Iran, 55,000; Iraq,* 27,000; Abu Phabi, 16,000; Libya,* 35,000; and Algeria,* 8,000.[8]

ARAB STRATEGY

While Russia has been active in the Middle East, the Arabs have not been idle, as the example of Libya will show. Libya's president, Colonel el-Qaddafi, has been wheeling and dealing with his country's annual 2.4 billion dollars earned from oil. Young, fiercely competitive, and a firm believer in Arab unity, Qaddafi has been "buying" alliances throughout Northern Africa and the Middle East. Consider the following:

(1) Uganda's army and air force were trained by Israeli military advisers. In addition, Israel has extended aid and credits to the tune of $25 million. A recent visit to Israel by Uganda's president, Major General Idi Amin, failed to raise an additional $10 million in aid. Shortly after, the general met with el-Qaddafi and proclaimed "the just struggle of the Palestinian people."[9] Following a reported promise of $26 million, Uganda broke relations with Israel, ordering its military advisers and embassy staff to leave the country.

A frightening possibility exists that because of oil the future will see a surge of anti-Israeli sentiment on an international scale. Congressman Wayne Aspinall, chairman of the powerful House Interior Committee, told *Time* magazine: "I am truly frightened by the potential conflict between pro-Israel sentiment in this country and our increasing reliance on Arab oil. I believe the U.S. is about to be caught in a Middle Eastern power play."[10]

(2) El-Qaddafi has called for the overthrow of the moderate King Hussein of Jordan and King Hassan of Morocco.

(3) Libya recently announced a restriction on the daily production of petroleum.[11] Is this an effort to put pressure on Western Europe to accept price increases? Is the decision based upon a desire to preserve the supply of this precious commodity so as to forestall the catastrophic political and economic consequences of one day running dry? Or is the objective to prevent other countries from boycotting Iraq oil in retaliation for the nationalization of the Iraq Petroleum Company? Whatever is behind it, clearly, the oil-rich countries will be increasingly emboldened to take such actions that are appropriate to their changed status.

(4) El-Qaddafi is providing arms and training for separatists in Chad and is supplying weapons to rebels in Ethiopia in an effort to overthrow Haile Selassie, whom he characterizes as "a lackey of

Israel."[12] In addition, he provides yearly grants of $125 million to Egypt and $45 million to Syria.

Libya's annual $2.4 billion in oil royalties will rise sharply in the next several years as Western Europe's oil needs increase. President Qaddafi's financial power is waxing, not waning. Potential Qaddafis in North Africa and the Middle East are watching the outcome.

ENERGY IMPORTATION AND OUR BALANCE OF PAYMENTS

Even if we survive with our Mideast oil supplies intact, what will our importation of oil do to our balance of payments? Balance of payments represent a complex inflow-outflow of money in four main categories: trade (imports-exports), tourism, investments, and military expenditures. Historically, a favorable balance of trade has primarily kept us in the black. However, in the past few years, even this advantage has begun to evaporate. In fact, in 1971, the United States experienced its first trade deficit in the century, $2.5 billion in contrast to a surplus of $2.1 billion in 1970.

Let us assume that by 1985 the price of gas will be about one dollar per 1,000 cubic feet and the price of oil about one dollar per barrel. This is actually close to the present domestic prices for industrial users; we are being conservative in our estimates. By 1985 our projected import needs of about 13 trillion cubic feet of gas and about 6 billion barrels of oil will have a minimum price tag of $19 billion a year!

The holders of these outflowing dollars don't have to purchase goods in America. They may elect to purchase their power plants, factories, and scientific equipment from high technology countries that have cheaper prices – such as Japan, England, France, West Germany, Israel, and Russia. America is precluded from a favorable trade balance because of high prices of domestically produced goods and services – and these could go higher by 1985, just when our need for a competitive capability will be at its zenith. The result of such strategy will be a United States industrial isolation from world markets.

We will be caught in a vicious circle: our country will soon pay a much higher price for energy, which will push up prices; meanwhile we will be importing energy and suffering a severe balance of trade and payments problem.

Not so for Russia. We have seen that Russia will be selling oil to Japan and gas to West Germany, France, and Italy. With income from the sale of energy, she will be in a strong balance-of-payments position.

It is interesting to note that the Ford Foundation recently funded a two-million-dollar study of national energy policy to be headed by S. David Freeman, former head of the Energy Policy Staff in the White House.[13] Included in the objectives of the study is an effort to determine the relationship between energy policy and the balance of payments dilemma.

NOTES

1. "Huge Siberian Gas Field Opens," *New York Times,* April 23, 1972.
2. "U.S.-Soviet Gas Deal?" *New York Times,* December 10, 1971, sect. 1, p. 8.
3. *Chemical and Engineering News* (May 29, 1972), p. 4.
4. "Tito Is Welcomed in Soviet Warmly," *New York Times,* June 6, 1972, p. 11.
5. "Soviet Announces Pact to Develop Libya's Oil Fields,"*New York Times,* March 5, 1972.
6. "The Croesus of Crisis," *Time* (April 10, 1972), p. 31.
7. "Iraq Takes Over Big Oil Company After Talks Fail," *New York Times,* June 2, 1972.
8. "The Arabs and Their Oil," *Chemical and Engineering News* (November 16, 1970), p. 62.
9. "The Croesus of Crisis," *Time,* p. 31.
10. Ibid.
11. *New York Times,* June 30, 1972.
12. "The Croesus of Crisis," *Time.*
13. *Chemical and Engineering News* (May 8, 1972), p. 24.

The Politics of Power — Domestic

AN ENERGY SHORTAGE DUE TO POLITICS?

There are some analysts, who, surveying the energy scene in this country, conclude that the energy crisis we have described in this book is artificial – a consequence of political, rather than physical, factors. For example:

(1) Passage of the Natural Gas Act of 1938. In 1938 Congress gave the Federal Power Commission (FPC) the authority to regulate the price of interstate shipments of gas. The Congress, on the consumer's behalf set low prices for gas. These low prices, the "energy watchers" reason, caused an adverse domino effect in our energy-producing industries.[1] The domino effect went like this: the electric utilities started switching from coal for energy to gas because it was cheap. This left the coal industry in a poor financial position; it left the coal industry with slight growth from 1945 until today. Meanwhile, the gas industry suffered from small profits; it did not earn the capital with which to explore for more gas fields. Both the coal and gas industries were hurt in terms of maintaining the profit incentives to explore for, and develop, our natural reserves.

(2) To protect domestic oil industries and in deference to the military, who feared dependence upon foreign energy sources, fed-

erally imposed import quotas have forced us to turn more to domestic sources than would have prevailed in a free market. We have only to abolish import quotas and the shortages will disappear.

(3) The utilities are responsible for the high consumption of electricity and natural gas as a direct result of extensive and intensive promotional campaigns to encourage the greater, and frequently extravagant, use of electricity and natural gas. We are all familiar with the slogan, "Keep a light turned on at night to discourage a burglar." Critics claim that predictions of greater energy consumption are self-fulfilling.

(4) The federal government has capitulated to the lunatic fringe in the environmental movement by permitting the delay in construction of necessary power systems – atomic power plants, oil pipelines, and the like – and prohibiting the exploration for energy resources in federally held lands.

(5) The American Petroleum Institute and the American Gas Association are the prime sources of information concerning our reserves of fossil fuels. Some energy watchers charge that they have conspired to underestimate our reserves in the hopes of obtaining greater tax relief and higher profits, which could then be used for further exploration.

Let us examine each of these arguments.

(1) Is the government to blame for the gas gap? Only to a slight degree. The fact is that today only 17 percent of gas consumption goes into the production of electricity. Gas consumption is on the rise for reasons other than just an increase in demand for electricity. It is being sought for heating purposes because it leaves no ashes to be transported; it is cheaper to "handle" than coal, in addition to being cheap by FPC fiat. Also, new air-pollution standards are accelerating the demand for gas because it is a cleaner-burning fuel than either coal or oil, by a wide margin.

Had the FPC never set artificially low prices for gas, our onrushing gas gap would be less severe. But by how much? As we have seen, the heart of the gas-gap problem is primarily a rising consumption rate rather than an FPC master bungling job.

Back in the 1940s when our gas consumption was rising about six percent per year, we had consumed less than ten percent of our gas heritage. Today our gas consumption rate is rising at about seven percent per year, but we now use over three times as much gas as twenty years ago.

DEMAND VS. SUPPLY FOR GAS[2]

	1950	1971
Rate of consumption in trillions of cubic feet per year	7	23
Rate of discovery of proved reserves in trillions of cubic feet per year	12	12

The impending gas gap is due partially to the FPC and partially to air-quality-standard upgrading; but it is mostly due to increasing millions of Americans who double their energy consumption every ten years.

Note in the chart above that we are presently consuming about twice as much gas per year as we are discovering. We have discussed the extraction problem as a resource becomes depleted: it becomes more time-consuming and more expensive to extract the last remaining quantities. This is our situation with oil and gas.

Moreover, the Federal Power Commission estimates that the rate at which we can produce natural gas from our reserves will peak in 1974 at about 24 to 25 trillion cubic feet.[3]

(2) Was the federal government wise in placing restrictions on the importation of foreign oil? The answer is damned if you do, damned if you don't. Eliminating shortages by abolishing quotas is nonsense.

Suppose that we had no oil import quotas? Then surely we would now be importing more than 22 percent of our oil from foreign sources, where it is cheaper to produce. But we would now be that much more vulnerable in the international strategy of brinkmanship, especially to oil-price changes or oil cutoffs for political reasons. We would have used less of our domestic reserves, but even these can't be utilized quickly in an emergency; it takes years to explore and develop oil fields.

Suppose that we had a 100 percent restriction against oil importation? We would now be further along than fifty percent in the depletion of our oil reserves, and we would be paying higher prices for gasoline. At the moment, however, we would be free of foreign dependency for oil – but we would be moving toward dependency at a vastly accelerated rate. Also, we might have developed a false sense of security about our true oil reserves with no strategy in mind to replace this vital resource.

(3) Are the utilities responsible for the high consumption of electrical energy and natural gas? We may as well ask which came first, the chicken or the egg? As we have seen, power consumption is directly related to a society's standard of living. The power companies could argue, with reason, that they have simply responded to the public's demand for greater goods and services. In doing so, they have stimulated the growth of whole new industries and their related employment opportunities: a host of home appliances, lighting for safety in high crime areas, for example. In short, energy deployment and economic growth are two sides of the same coin.

Their critics counter with the argument that the power companies have lured the public into wanting unnecessary luxuries of life. Surely both points of view are valid. However, if it is a Spartan way of life we must have, it should not be the responsibility of the utility companies to bring it about – just as they should not be credited with the aspirations of a society bent on improving its living standard.

Many of the amenities we have demanded extract a high energy cost. Consider the electrical requirements of the six largest household appliances:

HOUSEHOLD APPLIANCES AND POWER CONSUMPTION

	WATTS REQUIRED	KILOWATT-HOURS PER MONTH
Air conditioner (in season)	1,325	438
Electric range	12,000	100
Refrigerator	295	98
Clothes dryer	4,350	80
Window fan (in season)	200	58
Color television	315	38

Note that an air conditioner uses 4.38 times as much energy as an electric range; that an electric range uses about 2½ times the energy of a color television set.

In an era of growing energy shortages, all our power exercising devices will be open to questioning. Who is to decide which of these may be dispensed with? One man's amenity is another's necessity.

(4) The federal government has capitulated to no one. The Congress passed the National Environmental Policy Act (NEPA) of 1969 in response to a pressing need to do something about our rapidly deteriorating environment. This act requires that every federal agency authorized to license a new public works project study the potential environmental impact of the project and publish its findings along with a plan for reasonable alternatives to those projects. Although valid in concept, the law has opened up a veritable Pandora's box.

This law sets the stage for citizen groups to bring into court, and hence into delay, a myriad of public works plans, such as the siting of atomic power plants and the trans-Alaska pipeline among many others. All these plans have been stalled in the courts for years. It is unfortunate that many important projects have been "studied" in court for years when the nation was entitled to an evaluation in months.

For example, we have the problem of siting offshore atomic power plants. According to Ralph Lapp, the only logical solution to the "siting" problem of atomic power plants is to locate them offshore.[5] If we expect about one trillion watts of electricity in the future from atomic power plants of 1,000-megawatt capacity we shall need about 1,000 such plants. If located inland they would consume the lion's share of our fresh water, or they would require a 400-foot cooling tower for each plant.

However, if these atomic power plants were to be located some three miles offshore, then the "thermal pollution" problem is solvable. The power plants would "float" on platforms anchored behind wave breakers. The electricity generated would be transmitted to land by underwater cables. Each plant would release cooling water (sea water) at about 17 degrees above sea temperature – but the temperature at the periphery of a 5-acre area would be only 5 degrees hotter than ambient temperature. This could, indeed, stimulate fish growth; the thermal "pollution" could prove to be thermal "enrichment."

This design has been proposed by the Public Service and Gas Company of New Jersey. It is now before the Atomic Energy Commission for approval. Then, presumably, it faces court fights from environmental groups. The New Jersey Assembly voted 65 to 0 to withhold approval of this project. Apparently, the state legislators are susceptible to scare tactics just like anyone else: the safest possible atomic energy program frightens them more than what will happen in America during the energy-short decades. Similar proposals by Consolidated Edison to build offshore plants off Coney Island and Staten Island were stopped by public opposition.

The energy needs of our industrial nation and the need to preserve our natural environment will probably clash even harder in the coming years. This will be due to more than just strip-mining for coal at a rate of over tenfold compared to 1970, but due to the siting of synthetic fuel plants themselves. Can the conversion of coal into oil, gasoline, and gas be accomplished without causing more air pollution? If not, then the environmentalists will bring into court all plans for the siting and building of synthetic fuel plants, just as they have done with conventional oil refineries. John W. Sheihan, vice president of Shell Oil Co., has warned that the United States needs the equivalent of 58 new 160,000-barrel-per-day oil refineries by 1980 to meet our petroleum needs – assuming that the oil to be refined can be imported.[6] But only one is under construction; plans for the others are either in litigation or are being held up in anticipation of litigation.

Whether coal-conversion plants will be sufficiently pollution-free to satisfy the environmentalists and the courts remains to be seen. It is well known in air pollution abatement technology (applied to any source) that as you attempt to remove the remaining percentages of pollutants the costs rise exponentially. It may not be worth removing the last ten or five percent; if it must be worth it, then the consumer will pay higher prices for his synthetic fuels to the same degree.

One can expect that a synthetic fuel program will meet the same environmental opposition as have oil refineries and atomic power plants.

An example of the soaring costs for cleaning the last ten percent of auto emissions is presented below. It typifies the problem whether the source is a car or a power plant.

AIR POLLUTION ABATEMENT AND COSTS[7]

PERCENT REDUCTION	TOTAL COST PER CAR	COST PER INCREMENTAL PERCENT
62	$ 225	$ 4
92	600	12
96	1050	112

The litigation of organized environmentalists may well result in a public backlash once it is demonstrated that we are critically short of time to save our deteriorating energy supply systems. Protracted delays in implementing power procurement strategies on a crash-program basis will carry us to the brink of disaster.

This is not to say that the environmental groups are not right in their motives and ethics; it's their timing that's wrong.

Some critics have proposed that the federal government open publically held land for private exploration in search of energy resources. No situation better illustrates the contradictory pressure on the government. For example, one of the largest sources of low-sulfur coal is found on federally owned land in Utah and neighboring states.[8] In the past, the Bureau of Land Management of the Department of the Interior leased these lands to private companies for exploration and development. However, since February, 1971, it has refused to issue further licenses to develop these coal fields.

The desire to preserve the beauty of our land and the need for energy are on a collision course. In June, 1972, Governor Rockefeller vetoed a bill passed by the state assembly that would have banned oil and gas drilling around Long Island. He justified his action with the following words: "Because of our nation's growing energy needs it may be desirable at some future time to consider and permit drilling off the Atlantic coast of our state."[9]

(5) Some very highly placed energy watchers claim that the federal government can't possibly get accurate information about our fuel reserves because the only source for these estimates are private organizations such as the American Petroleum Institute and the American Gas Association. These organizations, the charge goes, have deliberately underestimated our reserves, hoping thereby to achieve greater incentives (profits and tax breaks) for exploration by fuel-producing industries. This charge appears incredible. How can collusion of any type create the fact that the amount of oil and gas discovered per foot of exploratory drilling is steadily declining? Is there a collusion to show exploration science as inept? The logical conclusion is that we are depleting the most readily available portion of our resources of gas and oil and are

now seeking the more inaccessible portions of these reserves.

The number of oil and gas wells drilled in the United States fell from 16,000 to 8,000 per year from 1956 to 1970.[10] However, the amount of oil and gas discovered per foot of drilling steadily declined, attesting to the scientific problems. It all means that the number of likely sites to sink a well are diminishing. We must drill deeper into the earth, into less permeable rock strata, and farther offshore in deeper water, and farther toward the Arctic ice cap to find recoverable oil and gas.

FEDERAL BUDGETING: STRATEGY FOR NONSURVIVAL

No one group or agency is to blame for our energy shortage. But there is, in fact, a villain in the piece – a lack of foresight, the same lack of foresight that is illustrated in the proposed federal budget for 1973.

The 1973 proposed budget for research and development is a strategy for nonsurvival. Of approximately $17,800 million, only $480 million is devoted to the research and development of "clean and abundant energy sources." That's less than three percent of our national research budget devoted to novel solutions to the energy crisis.

We will spend less than two-tenths of one percent (less than two cents out of every ten dollars!) of our total federal budget ($256 billion) on the research of problems pertaining to energy procurement.

Consider also that the total dollar flow in our economy directly derived from the energy industries is about $120 billion out of a $1.2 trillion economy – or about 10 percent of our GNP.[11] In addition, the price of energy is destined to rise dramatically, thereby making energy one of the largest items in our GNP.

THE BEST HOPE FOR POWER DEVELOPMENT: AN EXPANDED AEC

A major political battle now shaping up in Congress is the fight over just how the federal government should organize its research and development efforts in the field of energy. The major agencies now conducting research and development of power systems are the Atomic Energy Commission and the Department of the Interior; however, these are being joined by the Environmental Protection Agency, the National Science Foundation, the National Bureau of Standards, the Tennessee Valley Authority, and the Department of Housing and Urban Development.

Congress itself has no less than 13 committees that deal with energy!

No one superagency is responsible for energy development and deployment. There is no one guiding agency to give focus to the entire network of energy problems or perspective to any one energy research program. This chaos must end; we need a coherent national energy policy. Only an agency on the order of NASA or AEC can do this vital job.

The National Petroleum Council's Committee on the United States Energy Outlook (acting for the Secretary of the Interior), chaired by John G. McLean, has already published some of its findings in a first volume, which could be the basis for a strong first step toward establishing a national energy policy. Other volumes will follow. S. David Freeman, the former White House expert on energy, proposed a sound idea.[12] He believes that all research and development activities of the federal government should be consolidated under the Atomic Energy Commission (AEC) since they have the scientific personnel and equipment to handle the job – plus a proved record of capability.

Some experts propose the creation of a new superagency: an Energy Administration within a Department of Natural Resources.

Congress must decide how to organize our nation's energy research and development effort. If history is any guide, then perhaps the best proposal is to subsume all research and development under the AEC. This is also the recommendation of Glenn T. Seaborg, the world-famed authority on atomic energy and a former chairman of the AEC.

It might be worthwhile to see how our government previously gave birth to a new power industry (atomic energy) and then transferred that industry into the mainstream of our free-enterprise system. For those readers among the younger generation who may be unfamiliar with the origins and workings of the AEC, it might be useful to give a very brief account of the agency.

In 1942 the government collaborated with private industry in deep secret to create the massive technologies that led to the atomic bomb – code-named the Manhattan Project. By January of 1947 Congress created from the personnel and plants of the Manhattan Project the Atomic Energy Commission. By then a cumulative $2.2 billion had been spent for weapons development. From 1947 to 1953 the AEC researched the engineering requirements of nuclear power plants, all at government expense.

At that point the question arose whether the government should enter the energy business or whether private enterprise (the utilities and their subcontractors) should remain the energy producer for America. The answer came in the Atomic Energy Acts of 1953 and 1954: Congress restructured the AEC into a regulatory function as well as a development and operational function. The regulatory staff licenses utilities and industries that propose to build atomic power plants; hence the AEC Act of 1954 brought private industry into the atomic energy program. The development staff does research.

To date, the government has invested a cumulative $50 billion in atomic energy – weapon systems, power plants, and uranium enrichment facilities. Private industry has invested about $4 billion in power plants. However, by 1980 the cumulative investment by private industry in power plants will reach $50 billion. By the year 2000 the cumulative sum may reach $400 billion and an additional $60 billion per year thereafter for power plant construction and fuel costs combined.[13]

The only aspect of atomic power (for electricity) that is not in private hands is the fuel enrichment process. The government owns the facilities and charges for fuel enrichment. Even here plans are underway for a consortium of 20 corporations to build their own fuel enrichment processes with government aid.

When the AEC was born there was no private atomic energy program with which to come into conflict. The government, that is, the taxpayers, financed the birth of atomic power. The government then turned the information over to private enterprise.

Hopefully, this path will be open to us in our energy crisis. The research costs of fusion power, and the breeder reactor, and synthetic fuels will run into the hundreds of millions of dollars; but the development costs of building enough reactors (once researched) will run into the hundreds of billions of dollars – a thousand-fold difference. Recall that to close our gas gap of 1985 would require $32 billion for capital equipment, and about the same amount for the oil gap. To go beyond the gap and completely replace natural gas and oil by synthetic gas and oil in 1985 could reach $200 to $300 billion for capital equipment.

If the AEC becomes the USEC (United States Energy Commission) it could research the engineering requirements of new power systems and subsequently regulate the operation of these new systems by the private industries that would be licensed to use the new technologies. In the long run, the taxpayer would get his money's worth.

But can private industry raise its necessary capital without the disruption of our stock and bond markets? The total value of all stocks listed on the New York and American exchanges in 1971 was $750 billion. We've seen that just to close the gas and oil gaps of 1985 would require raising $60 to $70 billion over a 13-year period. What would be the capital outlays needed for a full nuclear age and all its electrical requirements? Obviously, a United States Energy Commission could research new power systems within the context of our free-enterprise system but the need is to finance the deployment of these new systems on a nationwide scale. Safeguards must be employed so that the government can participate in the solution to our energy shortage without becoming the owner of our power base.

NOTES

1. J. Eugene Cuccione, "The U.S. Power Crisis Has Arrived: Point of View," *New York Times,* May 7, 1972.
2. Ralph E. Lapp, "We're Running Out of Gas," *New York Times* Magazine, March 19, 1972, p. 34.
3. "Federal Power Commission Staff Report," *Chemical and Engineering News* (March 6, 1972), p. 4.
4. The Long Island Lighting Company and the Edison Electric Institute figures for a typical Long Island household.
5. "One Answer to the Atomic-Energy Puzzle – Put the Atomic Power Plants in the Ocean," *New York Times* Magazine, June 4, 1972, p. 30.
6. *Chemical and Engineering News* (May 29, 1972), p. 6.
7. "Cost vs. Benefits of Pollution Control," *Chemical and Engineering News* (June 19, 1972), p. 9.
8. William F. Reibenbacher, "The 'Energy' Crisis," *National Review* (June 9, 1972).
9. *Newsday,* June 10, 1972, p. 5.
10. American Petroleum Institute.
11. *Energy and Power, Scientific American* (September, 1971).
12. "Congress Probes U.S. Energy Research and Development Policy," *Chemical and Engineering News* (June 12, 1972).
13. Glenn T. Seaborg and William R. Corliss, *Man and Atom* (New York: E. P. Dutton and Co., 1971), p. 282.

Postscript

We have made progress in protecting the consumer, in legal safeguards for all citizens, in equal economic opportunities for the downtrodden, and in purging our environment of filth and exploitation.

But it is our environment that has most gripped our imagination. As the public's interest in the environment deepens, the issue of conservation becomes stronger. How many fields should be plowed? How many trees should be felled? How will we preserve endangered species from extinction? Where will we in the year 2000 find recreation, solitude, or that communion that can only come from contact with nature?

But these are not the issues that will make or break mankind. As important as they are, we need the perspective to see that they are secondary to the power that sustains our lives. If we can solve our energy problem we can win the time – and capacity – for the resolution of all our problems.

Recycling beer cans and old newspapers and conserving the Everglades should not be belittled, for they reveal the greatness of human concern and feeling, but they are not great issues: power procurement is an issue that can rescue or doom millions of people, and indeed, our future.

History implies that man, by his gift of creativity, is foresworn from a respectful status quo with nature. He is irretrievably committed to a perpetual series of rendezvous with some new destiny spawned by his indomitable curiosity.

Controlled nuclear fusion is an example. Future civilizations may hail it as man's greatest milestone in his quest for power, but its contribution may be only a brief notch in his erratic success story. Subsequently, after due creative agitation, man may yet approach a greater threshold – perhaps the transcendence of his intelligence, from a hydrocarbon life form to some new life form.

But whatever road man travels, every generation is both heir to all that came before and author of what comes after. In this sense, each generation is a trustee for humanity. The events of our own age invest this role with a special importance: with the acknowledgment that our power base is running low, that the fortunes of our institutions and our life-support system are inseparable, that the creation of controlled fusion would expand our technical capabilities to new horizons. History charges us with the gravest responsibility of any generation for designing decisions that will have the profoundest effect on the history of man.

An accurate appraisal of our heritage, an indefatigable concern for our posterity, the wisdom to choose the right options, these faculties raise us to the high level of a trustee for humanity. Can we fulfill this role? Will this generation sense its own historical influence and responsibilities?

WHAT CAN EACH OF US DO?

- Petition our local and national elected and appointed representatives to work for a meaningful overall energy program that should include developing controlled nuclear fusion on a first-priority, crash-program basis; development of the breeder reactor, but at a vastly accelerated pace; development of a synthetic fuel capacity to whatever degree is necessary to preclude overdependence on foreign sources and to tide us over to the nuclear age.

- Encourage the immediate development of a national center to research energy procurement and deployment, including the electrical needs of an atomic age.
- Encourage the establishment of a center to predict the economic, military, social, and political consequences of our impending power base transformations.
- Encourage the creation of a conference among our allies and trading partners to establish the means to share knowledge and energy resources in an orderly way.

Appendix

TIMETABLE FOR AN ENERGY CRISIS

Curtailed heating (and air conditioning) in shopping centers, department stores, theatres, and restaurants	the early 1970s
Curtailed heating (and air conditioning) in public buildings, schools, houses, and hospitals	the late 1970s
Gasoline rationing for everyone: no pleasure driving, car pools, increased public transportation	the late 1970s if we are obliged to supply Japan and Europe; otherwise, by the 1980s
Government control of key industries through the necessity of rationing energy	by the end of the 1970s
Government spending of hundreds of billions of dollars in crash programs for energy procurement	by the end of the 1970s

Strip-mining for coal on a vast scale in the U.S.; deep coal mining on a vast scale in most countries	1980
A "Great Depression" of 1929 scope: reduced building construction; reduced employment opportunities; a stock market collapse	by the 1980s if even one-half of the anticipated energy shortage materializes
Loss of world leadership to Russia due to a crippling energy shortage	by the 1980s
A world conflict over energy resources and possible military conflict	by the 1980s
Increased pollution due to coal consumption; a copper shortage for industrialized nations	1990
A water shortage in the U.S.	2000
Bringing in the nuclear age; can we meet the electrical requirements?	2010
World population reaches the limit of 10 billion; all usable land occupied (by farms or cities)	2020
A hotter world climate due to the "greenhouse effect" of increasing atmospheric carbon dioxide. Massive and unpredictable environmental consequences	2030

Index

(Numbers in italics refer to charts or graphs)

A

B

C